Cameos for Calculus
Visualization in the First-Year Course

© 2015 by the Mathematical Association of America, Inc.
Library of Congress Catalog Card Number 2015948230
Print edition ISBN 978-0-88385-788-5
Electronic edition ISBN 978-1-61444-120-5
Printed in the United States of America
Current Printing (last digit):
10 9 8 7 6 5 4 3 2

Cameos for Calculus

Visualization in the First-Year Course

Roger B. Nelsen
Lewis & Clark College

Published and Distributed by
The Mathematical Association of America

Council on Publications and Communications
Jennifer J. Quinn, *Chair*

Committee on Books
Fernando Gouvêa, *Chair*

Classroom Resource Materials Editorial Board
Susan G. Staples, *Editor*
Jennifer Bergner
Caren L. Diefenderfer
Christina Eubanks-Turner
Christopher Hallstrom
Cynthia J. Huffman
Brian Paul Katz
Paul R. Klingsberg
Brian Lins
Mary Eugenia Morley
Philip P. Mummert
Darryl Yong

CLASSROOM RESOURCE MATERIALS

Classroom Resource Materials is intended to provide supplementary classroom material for students—laboratory exercises, projects, historical information, textbooks with unusual approaches for presenting mathematical ideas, career information, etc.

101 Careers in Mathematics, 3rd edition edited by Andrew Sterrett

Archimedes: What Did He Do Besides Cry Eureka?, Sherman Stein

Arithmetic Wonderland, Andrew C. F. Liu

Calculus: An Active Approach with Projects, Stephen Hilbert, Diane Driscoll Schwartz, Stan Seltzer, John Maceli, and Eric Robinson

Calculus Mysteries and Thrillers, R. Grant Woods

Cameos for Calculus: Visualization in the First-Year Course, Roger B. Nelsen

Conjecture and Proof, Miklós Laczkovich

Counterexamples in Calculus, Sergiy Klymchuk

Creative Mathematics, H. S. Wall

Environmental Mathematics in the Classroom, edited by B. A. Fusaro and P. C. Kenschaft

Excursions in Classical Analysis: Pathways to Advanced Problem Solving and Undergraduate Research, by Hongwei Chen

Explorations in Complex Analysis, Michael A. Brilleslyper, Michael J. Dorff, Jane M. McDougall, James S. Rolf, Lisbeth E. Schaubroeck, Richard L. Stankewitz, and Kenneth Stephenson

Exploratory Examples for Real Analysis, Joanne E. Snow and Kirk E. Weller

Exploring Advanced Euclidean Geometry with GeoGebra, Gerard A. Venema

Game Theory Through Examples, Erich Prisner

Geometry From Africa: Mathematical and Educational Explorations, Paulus Gerdes

The Heart of Calculus: Explorations and Applications, Philip Anselone and John Lee

Historical Modules for the Teaching and Learning of Mathematics (CD), edited by Victor Katz and Karen Dee Michalowicz

Identification Numbers and Check Digit Schemes, Joseph Kirtland

Interdisciplinary Lively Application Projects, edited by Chris Arney

Inverse Problems: Activities for Undergraduates, Charles W. Groetsch

Keeping it R.E.A.L.: Research Experiences for All Learners, Carla D. Martin and Anthony Tongen

Laboratory Experiences in Group Theory, Ellen Maycock Parker

Learn from the Masters, Frank Swetz, John Fauvel, Otto Bekken, Bengt Johansson, and Victor Katz

Math Made Visual: Creating Images for Understanding Mathematics, Claudi Alsina and Roger B. Nelsen

Mathematics Galore!: The First Five Years of the St. Marks Institute of Mathematics, James Tanton

Methods for Euclidean Geometry, Owen Byer, Felix Lazebnik, and Deirdre L. Smeltzer

Ordinary Differential Equations: A Brief Eclectic Tour, David A. Sánchez

Oval Track and Other Permutation Puzzles, John O. Kiltinen

Paradoxes and Sophisms in Calculus, Sergiy Klymchuk and Susan Staples

A Primer of Abstract Mathematics, Robert B. Ash

Proofs Without Words, Roger B. Nelsen

Proofs Without Words II, Roger B. Nelsen

Rediscovering Mathematics: You Do the Math, Shai Simonson

She Does Math!, edited by Marla Parker

Solve This: Math Activities for Students and Clubs, James S. Tanton

Student Manual for Mathematics for Business Decisions Part 1: Probability and Simulation, David Williamson, Marilou Mendel, Julie Tarr, and Deborah Yoklic

Student Manual for Mathematics for Business Decisions Part 2: Calculus and Optimization, David Williamson, Marilou Mendel, Julie Tarr, and Deborah Yoklic

Teaching Statistics Using Baseball, Jim Albert

Visual Group Theory, Nathan C. Carter

Which Numbers are Real?, Michael Henle

Writing Projects for Mathematics Courses: Crushed Clowns, Cars, and Coffee to Go, Annalisa Crannell, Gavin LaRose, Thomas Ratliff, and Elyn Rykken

MAA Service Center
P.O. Box 91112
Washington, DC 20090-1112
1-800-331-1MAA   FAX: 1-301-206-9789

*Dedicated with love to my sister Kathleen —
who never studied calculus and yet is
one of the most creative persons I know.*

Preface

> *A dull proof can be supplemented by a geometric analogue so simple and beautiful that the truth of a theorem is almost seen at a glance.*
>
> Martin Gardner
>
> *Behold!*
>
> Bhāskara

A thespian or cinematographer might define a *cameo* as "a brief appearance of a known figure," while a gemologist or lapidary might define it as "a precious or semiprecious stone." How might a mathematician define it? In this book I present fifty short enhancements for the first-year calculus course in which a geometric figure briefly appears, which I call *Cameos for Calculus*. Some of the Cameos illustrate mainstream topics such as the derivative (Cameo 3), combinatorial formulas used to compute Riemann sums (Cameo 16), or the geometry behind many geometric series (Cameo 33). Other Cameos present topics accessible to students at the calculus level but not usually encountered in the course, such as the Cauchy-Schwarz inequality (Cameo 24), the arithmetic mean-geometric mean inequality (Cameos 10, 15, and 45), and the Euler-Mascheroni constant (Cameo 37).

In an early 1990s article "Visual Thinking in Calculus" (in *Visualization in Teaching and Learning Mathematics*, W. Zimmerman and S. Cunningham, editors, MAA, 1991), Walter Zimmerman wrote:

> Of all undergraduate mathematics courses, none offers more interesting and varied opportunities for visualization than calculus. Most of the concepts and many problems of calculus can be represented graphically. Recognizing the importance of graphics in calculus, texts are adorned by numerous figures and diagrams. In many cases, however, these are little more than decorations. In selected cases, diagrams may be used directly as a tool in problem solving, but considering the calculus course as a whole, geometrical reasoning is used inconsistently at best, and the role of visual thinking is not seriously addressed.

Many of the Cameos are adapted from articles published in journals of the MAA, such as the *American Mathematical Monthly*, *Mathematics Magazine*, and the *College Mathematics Journal*. Some come from other mathematical journals, and some were created for this book. By gathering the Cameos into a book I hope that they will be more accessible to teachers of calculus, both for use in the classroom and as supplementary explorations for students.

There are fifty Cameos in the book, grouped into five sections: Part I, Limits and Differentiation; Part II, Integration; Part III, Infinite Series; Part IV, Additional Topics; and Part V, Appendix: Some Precalculus Topics. Many of the Cameos include exercises, so Solutions to the Exercises follow Part V. The book concludes with References and an Index.

Thanks to Susan Staples and the members of the editorial board of the Classroom Resource Materials book series for their careful reading of an earlier draft of the book and for their many helpful suggestions. I would also like to thank Carol Baxter, Associate Director of Publications; Beverly Ruedi, Electronic Production and Publishing Manager; Samantha Webb, Marketing Production Assistant; and Stephen Kennedy, Senior Acquisitions Editor of the MAA's publication staff for their expertise in preparing this book for publication. Finally, special thanks go to many fellow teachers in the calculus community for encouraging me to work on this publication.

Roger B. Nelsen
Lewis & Clark College
Portland, Oregon

Contents

Preface	ix
Part I Limits and Differentiation	**1**
1 The limit of $(\sin t)/t$	3
2 Approximating π with the limit of $(\sin t)/t$	5
3 Visualizing the derivative	7
4 The product rule	9
5 The quotient rule	11
6 The chain rule	13
7 The derivative of the sine	15
8 The derivative of the arctangent	17
9 The derivative of the arcsine	19
10 Means and the mean value theorem	21
11 Tangent line inequalities	23
12 A geometric illustration of the limit for e	27
13 Which is larger, e^π or π^e? a^b or b^a?	29
14 Derivatives of area and volume	31
15 Means and optimization	33
Part II Integration	**39**
16 Combinatorial identities for Riemann sums	41
17 Summation by parts	47
18 Integration by parts	51
19 The world's sneakiest substitution	53
20 Symmetry and integration	57
21 Napier's inequality and the limit for e	61

22	The nth root of $n!$ and another limit for e	65
23	Does shell volume equal disk volume?	67
24	Solids of revolution and the Cauchy-Schwarz inequality	71
25	The midpoint rule is better than the trapezoidal rule	75
26	Can the midpoint rule be improved?	77
27	Why is Simpson's rule exact for cubics?	79
28	Approximating π with integration	81
29	The Hermite-Hadamard inequality	83
30	Polar area and Cartesian area	87
31	Polar area as a source of antiderivatives	89
32	The prismoidal formula	91

Part III Infinite Series — 93

33	The geometry of geometric series	95
34	Geometric differentiation of geometric series	99
35	Illustrating a telescoping series	101
36	Illustrating applications of the monotone sequence theorem	103
37	The harmonic series and the Euler-Mascheroni constant	107
38	The alternating harmonic series	111
39	The alternating series test	113
40	Approximating π with Maclaurin series	115

Part IV Additional Topics — 119

41	The hyperbolic functions I: Definitions	121
42	The hyperbolic functions II: Are they circular?	125
43	The conic sections	129
44	The conic sections revisited	133
45	The AM-GM inequality for n positive numbers	135

Part V Appendix: Some Precalculus Topics — 139

46	Are all parabolas similar?	141
47	Basic trigonometric identities	143

48	**The addition formulas for the sine and cosine**	**145**
49	**The double angle formulas**	**147**
50	**Completing the square**	**149**
	Solutions to the Exercises	**151**
	References	**163**
	Index	**167**
	About the Author	**171**

PART I
Limits and Differentiation

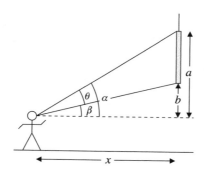

CAMEO 1

The limit of $(\sin t)/t$

Almost all modern calculus texts include a proof that $\lim_{t\to 0} (\sin t)/t = 1$. Many do so in the same way: for t in $(0, \pi/2)$ compare the areas of the circular sector and two triangles illustrated in Figure 1.1a. The area of $\triangle OAP$ is $(\sin t)/2$, the area of sector AOP is $t/2$, the area of $\triangle OAB$ is $(\tan t)/2$, and thus $(\sin t)/2 \le t/2 \le (\tan t)/2$. These inequalities are then cleverly manipulated to obtain $\cos t \le (\sin t)/t \le 1$, from which it follows that $\lim_{t\to 0^+} (\sin t)/t = 1$ (the limit as $t \to 0^-$ is usually obtained by noting that $(\sin(-t))/(-t) = (\sin t)/t)$. It is the manipulation of the inequalities that may cause some students difficulty.

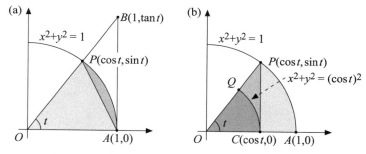

Figure 1.1. Two illustrations for the limit of $(\sin t)/t$

The manipulation of the inequalities can be simplified by using *one* triangle and *two* circular sectors, as illustrated in Figure 1.1b. The area of sector COQ is $t(\cos t)^2/2$, the area of $\triangle OCP$ is $(\sin t)(\cos t)/2$, the area of sector AOP is $t/2$, and hence $t(\cos t)^2/2 \le (\sin t)(\cos t)/2 \le t/2$. Multiplication by $2/(t\cos t)$ immediately yields $\cos t \le (\sin t)/t \le 1/\cos t$. The limit now follows from the squeeze theorem once we show that $\lim_{t\to 0} \cos t = 1$.

To do so, see Figure 1.2.

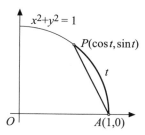

Figure 1.2. An illustration for two cosine limits

For t in $(-\pi/2, 0) \cup (0, \pi/2)$, $\cos t < 1$ and the length of chord AP is less than the length t of arc AP, so that
$$\sqrt{(\cos t - 1)^2 + \sin^2 t} < |t| \text{ or } 2 - 2\cos t < t^2.$$
Thus $1 - (t^2/2) < \cos t < 1$ and $\lim_{t \to 0} \cos t = 1$ follows from the squeeze theorem.

Exercise 1.1. Show that $\lim_{t \to 0} (\tan t)/t = 1$.

Exercise 1.2. Show that $\lim_{t \to 0} (1 - \cos t)/t = 0$. (Hint: use Figure 1.2.)

The expression in the title of this Cameo and its limit at 0 together define the *sinc* function (from the Latin *sinus cardinalis*), which finds applications in Fourier analysis and signal processing:
$$\operatorname{sinc}(x) = \begin{cases} (\sin x)/x, & x \neq 0 \\ 1, & x = 0. \end{cases}$$

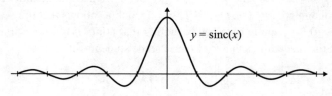

Figure 1.3. $y = \operatorname{sinc}(x)$ for x in $[-5\pi, 5\pi]$

SOURCE: Adapted from C. Alsina and RBN, "Teaching tip: The limit of $(\sin t)/t$," *College Mathematics Journal*, **41** (2010), p. 192.

CAMEO 2

Approximating π with the limit of $(\sin t)/t$

The primary, and often only, application of the limit in the preceding Cameo in many calculus texts is differentiating the sine and cosine functions. Few texts provide another application of the limit. Here is a simple one, approximating the value of π.

After the limit has been established, set $t = \pi/n$ for large n, so that $(\sin(\pi/n))/(\pi/n) \approx 1$, or equivalently, $n \sin(\pi/n) \approx \pi$. Of course, we need to evaluate $n \sin(\pi/n)$ without using π, so we begin with $2 \sin(\pi/2) = 2$, and then employ the half angle formulas for the sine and cosine to compute $n \sin(\pi/n)$ where n is a power of 2.

The half angle formulas (see Cameo 49) are

$$\sin \frac{x}{2} = \sqrt{\frac{1 - \cos x}{2}} \quad \text{and} \quad \cos \frac{x}{2} = \sqrt{\frac{1 + \cos x}{2}}.$$

We arrange our work in the table below, computing each row (after the $n = 2$ row) from the previous row:

Table 2.1. Approximating π

n	$\sin(\pi/n)$	$\cos(\pi/n)$	$n \sin(\pi/n)$
2	1.0	0.0	2.0
4	0.7071067812	0.7071067812	2.8284271248
8	0.3826834324	0.9238795325	3.0614674589
16	0.1950903220	0.9807852804	3.1214451522
32	0.0980171403	0.9951847267	3.1365484904
64	0.0490676743	0.9987954562	3.1403311573
128	0.0245412285	0.9996988187	3.1412772512
256	0.0122715383	0.9999247018	3.1415138074
512	0.0061358846	0.9999811753	3.1415729225
1024	0.0030679567	0.9999952938	3.1415876920

There is an ancient geometric interpretation of the results in the $n \sin(\pi/n)$ column, one that explains why the approximations approach π through values less than π. With Figure 2.1 we can show that $n \sin(\pi/n)$ equals the area of a regular polygon with $2n$ sides (a "$2n$-gon") for $n \geq 2$ inscribed in a circle of radius 1.

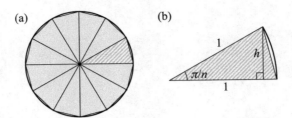

Figure 2.1. Inscribing a $2n$-gon in a circle of radius 1

The $2n$-gon in Figure 2.1a has been partitioned into $2n$ isosceles triangles, each with a vertex angle at the center of the circle measuring $2\pi/2n = \pi/n$ radians. The altitude h of one of these triangles (in Figure 2.1b) is $h = \sin(\pi/n)$, hence the area is $(1/2)\sin(\pi/n)$. Since there are $2n$ triangles, the area of the inscribed $2n$-gon is $n\sin(\pi/n)$, and is less than the area π of the circle.

Exercise 2.1. Show that for $n > 2$ the results in the $n\sin(\pi/n)$ column of Table 2.1 equal one-half the circumference of a regular n-gon similarly inscribed in a circle of radius 1.

Exercise 2.2. Use the result of Exercise 1.1 ($\lim_{x\to 0}(\tan x)/x = 1$) to generate another sequence of approximations to π. Is there a geometric interpretation of this sequence? (Hint: consider the areas of regular n-gons circumscribed about a circle of radius 1.)

In Cameos 28 and 40 we explore additional ways to approximate π using calculus.

CAMEO 3

Visualizing the derivative

Perhaps the most common way of introducing the derivative of a function f is to consider the problem of finding the slope of the tangent line to graph of $y = f(x)$ at a point $P = (a, f(a))$. This is usually accomplished by choosing a variable point $Q = (x, f(x))$, computing the slope m_{PQ} of the secant PQ, and letting Q approach P.

In this Cameo we show that with the addition of another vertical axis (which we call the *slope axis*, or *m-axis*), we can see the slopes of the secants converge. Since the slope of a line corresponds to the vertical displacement resulting from a horizontal displacement of one unit, we draw the m-axis as the vertical line $x = a + 1$ with its origin at $(a + 1, f(a))$ and the same scale as the y-axis. See Figure 3.1a.

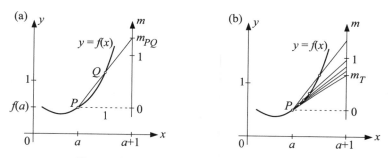

Figure 3.1. A method to visualize the derivative

Consequently, when we extend the secant PQ to intersect the m-axis, the point of intersection has the coordinate m_{PQ} on the m-axis, as illustrated in Figure 1.1a. As we move Q towards P, the secant lines converge to the tangent line, and the slopes m_{PQ} converge to the slope m_T of the tangent line on the m-axis, as illustrated in Figure 3.1b. In this example, it appears that m_T may be about $2/3$, and now the exact value of m_T can be found using the limit process with the difference quotient for m_{PQ}.

SOURCE: N. A. Friedman, "A picture for the derivative," *American Mathematical Monthly*, **84** (1977), pp. 470–471.

CAMEO 4
The product rule

The standard approach for deriving the product rule starts with the definition of the derivative, which has the obvious advantage of reinforcing the definition. However, some students find little that is intuitive or inspiring in that approach, since they see the algebra employed as tricky at best and totally confusing at worst. For such students a second approach may be beneficial.

In this approach we use a technique common in mathematics: we prove a special case first, and then use it to prove the general case. Our special case of a product is the *square of a function*, since every two-term product can be written as a difference of squares. See Figure 4.1.

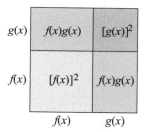

Figure 4.1. Expressing a product in terms of squares

Since
$$[f(x) + g(x)]^2 = [f(x)]^2 + 2f(x)g(x) + [g(x)]^2$$
we have
$$f(x)g(x) = \frac{1}{2}\left([f(x) + g(x)]^2 - [f(x)]^2 - [g(x)]^2\right). \tag{4.1}$$

We now derive the formula for the derivative of the square of a differentiable function *without* using the power, product, or chain rules. By definition we have
$$\frac{d}{dx}[f(x)]^2 = \lim_{\Delta x \to 0} \frac{[f(x+\Delta x)]^2 - [f(x)]^2}{\Delta x}$$
when the limit exists. The difference quotient in the limit can be written as
$$\frac{[f(x+\Delta x)]^2 - [f(x)]^2}{\Delta x} = [f(x+\Delta x) + f(x)] \cdot \frac{[f(x+\Delta x)] - [f(x)]}{\Delta x}$$

9

and consequently, if $f(x)$ is differentiable then so is $[f(x)]^2$ and

$$\frac{d}{dx}[f(x)]^2 = \lim_{\Delta x \to 0}[f(x + \Delta x) + f(x)] \cdot \frac{[f(x + \Delta x)] - [f(x)]}{\Delta x}$$

$$= \lim_{\Delta x \to 0}[f(x + \Delta x) + f(x)] \cdot \lim_{\Delta x \to 0}\frac{[f(x + \Delta x)] - [f(x)]}{\Delta x}.$$

Hence

$$\boxed{\frac{d}{dx}[f(x)]^2 = 2f(x)f'(x).} \tag{4.2}$$

Exercise 4.1. In the above derivation we use the fact that $\lim_{\Delta x \to 0} f(x + \Delta x) = f(x)$. Why is this true?

Exercise 4.2. Use (4.1) and (4.2) to derive the *product rule*: If $f(x)$ and $g(x)$ are differentiable, then so is $f(x)g(x)$ and

$$\boxed{\frac{d}{dx}f(x)g(x) = f(x)g'(x) + g(x)f'(x).} \tag{4.3}$$

SOURCE: R. Euler, "A note on differentiation," *College Mathematics Journal*, **17** (1986), pp. 166–167.

CAMEO 5

The quotient rule

In Cameo 4, we derived the product rule for differentiation by first considering a special case, the derivative of the square of a function. Can we do something similar with the quotient rule?

For the quotient rule, the special case we consider is sometimes called the *reciprocal rule*: if g is differentiable and $g(t) \neq 0$, then

$$\boxed{\frac{d}{dt}\left(\frac{1}{g(t)}\right) = \frac{-g'(t)}{[g(t)]^2}}, \tag{5.1}$$

which we can combine with the product rule to differentiate a general quotient:

$$\frac{d}{dt}\left(\frac{f(t)}{g(t)}\right) = \frac{d}{dt}\left(f(t) \cdot \frac{1}{g(t)}\right). \tag{5.2}$$

A simple quotient with which students are familiar is the slope of a line or line segment. Consider a line segment OP from the origin O to $P = (g(t), 1)$ with a variable slope given by $m(t) = 1/g(t)$ at time t as shown in Figure 5.1, where the "rise" is 1 and the "run" $g(t)$ is a nonzero differentiable function.

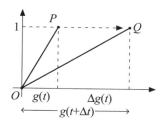

Figure 5.1. Illustrating the reciprocal rule

Suppose that the endpoint P moves to $Q = (g(t + \Delta t), 1)$ at time $t + \Delta t$ so that the slope of OQ is $m(t + \Delta t) = 1/g(t + \Delta t)$. Now consider the resulting change in slope $\Delta m(t) = m(t + \Delta t) - m(t)$ as P moves to Q:

$$\Delta m(t) = \frac{1}{g(t + \Delta t)} - \frac{1}{g(t)} = \frac{g(t) - [g(t) + \Delta g(t)]}{g(t)g(t + \Delta t)}$$

$$= \frac{-\Delta g(t)}{g(t)g(t + \Delta t)}.$$

Now divide both sides by Δt and take the limit as $\Delta t \to 0$:

$$m'(t) = \lim_{\Delta t \to 0} \frac{\Delta m(t)}{\Delta t} = \lim_{\Delta t \to 0} \frac{-\frac{\Delta g(t)}{\Delta t}}{g(t)g(t+\Delta t)} = \frac{-g'(t)}{[g(t)]^2},$$

which, since $m(t) = 1/g(t)$, establishes (5.1).

Exercise 5.1. In the above derivation we use the fact that $\lim_{\Delta t \to 0} g(t + \Delta t) = g(t)$. Why is this true?

Exercise 5.2. Use (5.1) and (5.2) to derive the *quotient rule*: if $f(t)$ and $g(t)$ are differentiable and $g(t) \neq 0$, then $f(t)/g(t)$ is differentiable and

$$\boxed{\frac{d}{dt}\left(\frac{f(t)}{g(t)}\right) = \frac{g(t)f'(t) - f(t)g'(t)}{[g(t)]^2},} \quad (5.3)$$

or, replacing $f(t)$ and $g(t)$ by u and v respectively, $(u/v)' = (vu' - uv')/v^2$.

Exercise 5.3. Here is another "proof" of the quotient rule using the product rule. Set $y = u/v$ so that $yv = u$, then product rule yields $yv' + y'v = u'$. Solving for y' yields $y' = (u/v)' = (vu' - uv')/v^2$. But there is a serious flaw in this "proof." What is it?

CAMEO 6

The chain rule

In order to illustrate the chain rule geometrically, we begin with a way to graph a composite function $y = f(g(x))$ from the graphs of two functions $y = f(x)$ and $y = g(x)$ (when, of course, the range of g is a subset of the domain of f). One way to do it is illustrated in Figure 6.1.

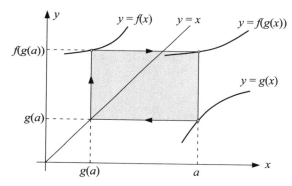

Figure 6.1. Graphing a composite function

Start with a point a on the x-axis, and use the graph of $y = g(x)$ to locate $g(a)$ on the y-axis. Then by means of the graph of $y = x$ we can relocate $g(a)$ on the x-axis. Now use the graph of $y = f(x)$ to locate $f(g(a))$ on the y-axis, and translate this value horizontally to lie above $(a, 0)$, and thus we have the point $(a, f(g(a)))$ on the graph of $y = f(g(x))$ as desired.

As Figure 6.1 illustrates, there is a moving frame (shaded gray) with one corner on the diagonal $y = x$, the two adjacent corners on the graphs of f and g, and the opposite corner on the graph of $f \circ g$.

To illustrate the chain rule, let f and g be differentiable, choose another point t on the x-axis close to a, and repeat the procedure to locate $(g(t), f(g(t)))$ on the graph of $y = f(x)$ and $(t, f(g(t)))$ on the graph of $y = f(g(x))$. Then draw the three secant lines, the heavy dashed segments shown in Figure 6.2.

We now compute the slopes of the three secant lines:

The slope of the secant of g between a and t is $\frac{g(t)-g(a)}{t-a}$.

The slope of the secant of f between $g(a)$ and $g(t)$ is $\frac{f(g(t))-f(g(a))}{g(t)-g(a)}$.

The slope of the secant of $f \circ g$ between a and t is $\frac{f(g(t))-f(g(a))}{t-a}$.

13

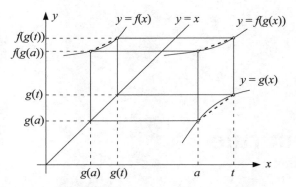

Figure 6.2. Illustrating the chain rule

Clearly the third slope is the product of the first two. In the limit as t approaches a, the three secant lines become tangent lines with slopes $g'(a)$, $f'(g(a))$, and $(f \circ g)'(a)$, respectively, thus

$$\boxed{(f \circ g)'(a) = f'(g(a)) \cdot g'(a).}$$

A minor modification of the illustration is needed if the secant of g between a and t is horizontal. In this case there is just one line through the two points on the graph of g, yielding a single point on the graph of f rather than a secant. It then follows that the secant of $f \circ g$ is also horizontal.

SOURCE: Adapted from Karl Menger's 1955 text *Calculus: A Modern Approach*, recently reprinted by Dover.

CAMEO 7

The derivative of the sine

Most textbooks prove that the derivative of the sine is the cosine using the definition of the derivative, the addition formula for the sine, and two limits previously derived. Here is a visual plausibility argument for this derivative that may help students see that the derivative of the sine should be the cosine.

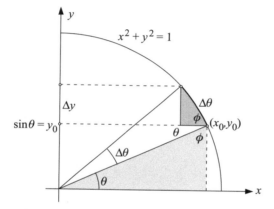

Figure 7.1. Illustrating the derivative of the sine

The smaller dark gray region is not a triangle, but approaches one similar to the light gray triangle in the limit as $\Delta\theta \to 0$, so that

$$\frac{dy}{d\theta} \approx \frac{\Delta y}{\Delta \theta} = \frac{x_0}{1} = \sin\phi = \cos\theta.$$

Exercise 7.1. Illustrate the derivative of the cosine in a similar manner.

SOURCE: D. Hartig, "On the differentiation formula for $\sin\theta$," *American Mathematical Monthly*, **96** (1989), p. 252, and S. Sridharma, "The derivative of $\sin\theta$," *College Mathematics Journal*, **30** (1999), pp. 314–315.

CAMEO 8

The derivative of the arctangent

The simplest of the derivatives of the inverse trigonometric functions is the derivative of the arctangent. A purely geometric derivation of the derivative of the arctangent can be obtained from the formula for the area $(1/2)r^2\theta$ of a sector of a circle of radius r whose angle at the origin is θ (see Figure 8.1a).

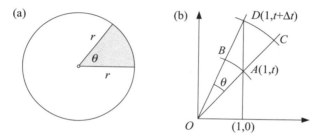

Figure 8.1. The derivative of the arctangent

The arctangent is an odd function (since the tangent is), so its derivative is even. Hence we need only find $d\arctan t/dt$ for $t \geq 0$. In Figure 8.1b $\widehat{AB}$ is an arc of a circle with radius $\sqrt{1+t^2}$, and $\widehat{CD}$ is an arc of a circle with radius $\sqrt{1+(t+\Delta t)^2}$ for $\Delta t > 0$ (the case where $\Delta t < 0$ is similar). Hence $\theta = \arctan(t + \Delta t) - \arctan t$. Since the area of sector AOB is less than the area $\Delta t/2$ of triangle OAD, we have

$$\frac{1}{2}(1+t^2)[\arctan(t+\Delta t) - \arctan t] < \frac{\Delta t}{2}$$

or

$$\frac{\arctan(t+\Delta t) - \arctan t}{\Delta t} < \frac{1}{1+t^2}.$$

Similarly the area of triangle OAD is less than the area of sector COD so that

$$\frac{\Delta t}{2} < \frac{1}{2}[1+(t+\Delta t)^2][\arctan(t+\Delta t) - \arctan t]$$

or

$$\frac{1}{1+(t+\Delta t)^2} < \frac{\arctan(t+\Delta t) - \arctan t}{\Delta t},$$

and therefore

$$\frac{1}{1+(t+\Delta t)^2} < \frac{\arctan(t+\Delta t) - \arctan t}{\Delta t} < \frac{1}{1+t^2}.$$

(When $\Delta t < 0$ we obtain the same terms but with the inequality signs reversed.)

Thus from the squeeze theorem for limits

$$\lim_{\Delta t \to 0^+} \frac{\arctan(t+\Delta t) - \arctan t}{\Delta t} = \frac{1}{1+t^2},$$

and similarly for the limit as $\Delta t \to 0^-$. So the ordinary limit as $\Delta t \to 0$ exists, and we have

$$\frac{d}{dt} \arctan t = \frac{1}{1+t^2}.$$

The derivatives of other inverse trigonometric functions now follow using the chain rule and identities:

$$\arcsin t = \arctan \frac{t}{\sqrt{1-t^2}}, \quad \arccos t = \frac{\pi}{2} - \arcsin t, \quad \text{arcsec}\, t = \arccos \frac{1}{t}, \quad \text{etc.}$$

Exercise 8.1. Follow the same procedure using Figure 8.2 to show that $\frac{d}{dx} \tan \theta = \sec^2 \theta$. (Hint: $|OA| = \sec \theta$, $|OD| = \sec(\theta + \Delta\theta)$.)

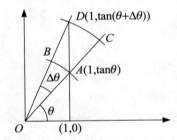

Figure 8.2. The derivative of the tangent

SOURCE: N. Schaumberger, "The derivatives of arcsec x, arctan x, and tan x," *College Mathematics Journal*, **17** (1986), pp. 244–246.

CAMEO 9

The derivative of the arcsine

Most calculus texts use implicit differentiation to evaluate the derivatives of the inverse trigonometric functions—two or three as examples and the others as exercises. With the fact that the area of a sector of the unit circle equals one-half the radian measure of the angle at the origin, an area integral, and the fundamental theorem, we have the following direct derivation of the derivative of the inverse sine.

As with the arctangent in the preceding Cameo, the arcsine is an odd function (since the sine is), so its derivative is even. Hence we need only find $d \sin^{-1} x/dx$ for $x \in [0, 1)$. See Figure 9.1.

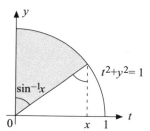

Figure 9.1. The derivative of the arcsine

The area of the shaded region can be computed in two ways:

$$\frac{1}{2}\sin^{-1} x = \int_0^x \sqrt{1-t^2}\,dt - \frac{1}{2}x\sqrt{1-x^2}.$$

Multiplication by 2 and differentiation yields

$$\frac{d}{dx}\sin^{-1} x = 2\sqrt{1-x^2} - \left(\sqrt{1-x^2} - \frac{x^2}{\sqrt{1-x^2}}\right) = \frac{1}{\sqrt{1-x^2}}.$$

SOURCE: Adapted from M. R. Spiegel, "On the derivatives of trigonometric functions," *American Mathematical Monthly*, **63** (1956), pp. 118–120.

CAMEO 10

Means and the mean value theorem

The *mean value theorem* is a staple of every calculus course. However, textbooks rarely explain the relationship between this theorem and *means*. The word "mean" comes from the French *moyen*, "middle," "medium," or "average." A mean of two numbers a and b (which we will take to be real and positive) is a number c that lies between a and b. Familiar examples are the *arithmetic mean* $(a+b)/2$ (the usual method of averaging numbers such as exam scores) and the *geometric mean* $\sqrt{ab}$. The origin of the geometric mean is, of course geometry—it is the side length of a square with the same area as an a-by-b rectangle.

The mean value theorem ("MVT") states that if f is a function continuous on $[a,b]$ and differentiable on (a,b), then there exists a number c in (a,b) such that

$$\frac{f(b)-f(a)}{b-a} = f'(c). \tag{10.1}$$

The relevant portion of the theorem for this Cameo is the fact that c is in (a,b), and hence c is a mean of a and b. All we need to do is solve (10.1) for c, the number in (a,b) where the tangent line is parallel to the secant line joining $(a, f(a))$ and $(b, f(b))$.

Example 10.1. Let $f(x) = x^2$ on $[a,b]$. The MVT yields $(b^2 - a^2)/(b-a) = 2c$, so that $c = (a+b)/2$, the arithmetic mean of a and b. See Figure 10.1.

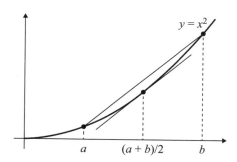

Figure 10.1. The MVT for $f(x) = x^2$ on $[a,b]$

Example 10.2. Let $f(x) = 1/x$ on $[a,b]$. The MVT yields $[(1/b)-(1/a)]/(b-a) = -1/c^2$, from which we have $c = \sqrt{ab}$, the geometric mean of a and b. See Figure 10.2.

21

CAMEO 10. Means and the mean value theorem

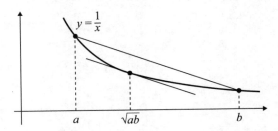

Figure 10.2. The MVT for $f(x) = 1/x$ on $[a, b]$

Exercise 10.1. Use the MVT to find the mean c for the function $y = \ln x$ on $[a, b]$. This mean is naturally called the *logarithmic mean* of a and b. See Figure 10.3.

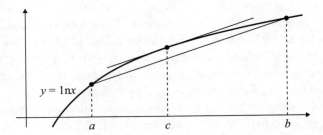

Figure 10.3. The MVT for $f(x) = \ln x$ on $[a, b]$

Exercise 10.2. An unusual mean of two positive numbers a and b is the *identric mean*, $(1/e) \cdot (b^b/a^a)^{1/(b-a)}$. Derive this mean using the MVT with $f(x) = x \ln x$ on $[a, b]$.

So, why are means useful in a calculus course? See Cameo 15 for an application of the arithmetic and geometric means to solve some optimization problems where first-year calculus techniques fail, and Cameo 29 for an application of the logarithmic mean to deriving inequalities for the number e.

We study relationships among the various means in Cameos 11, 12, 15, 21, and 29.

CAMEO 11

Tangent line inequalities

Tangent lines to the graph of a concave up function lie below the graph of the function, and tangent lines to the graph of a concave down function lie above the graph of the function. This simple observation can be exploited to derive a variety of inequalities.

Example 11.1. Is $(\pi/e) + (e/\pi) > 2$? One way to answer the question is to use a calculator. Another way is to show that *the sum of a positive number and its reciprocal is always at least 2*, i.e.,

$$\text{if } x > 0 \text{ then } x + \frac{1}{x} \geq 2. \tag{11.1}$$

with equality if and only if $x = 1$.

Consider the graph of $y = 1/x$ and its tangent line $y = 2 - x$ at (1,1), as shown in Figure 11.1. Since the curve is concave up, $1/x \geq 2 - x$, and (11.1) follows.

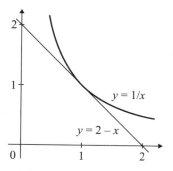

Figure 11.1. Graphs of $y = 1/x$ and $y = 2 - x$

Example 11.2. *The AM-GM inequality for two numbers.* We encountered the arithmetic mean $(a+b)/2$ and the geometric mean $\sqrt{ab}$ for positive numbers a and b in the preceding Cameo. The AM-GM (for arithmetic mean-geometric mean) inequality states that

$$\text{if } a \text{ and } b \text{ are positive, then } \frac{a+b}{2} \geq \sqrt{ab} \tag{11.2}$$

with equality if and only if $a = b$. To prove (11.2), we first show that $(1+x)/2 \geq \sqrt{x}$ for $x > 0$. See Figure 11.2. Since the graph of $y = \sqrt{x}$ is concave down, the line $y = (1+x)/2$

tangent to the curve at (1,1) lies above it, and $(1+x)/2 \geq \sqrt{x}$ follows (with equality if and only if $x = 1$).

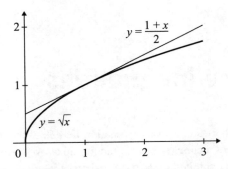

Figure 11.2. Graphs of $y = \sqrt{x}$ and $y = (1+x)/2$

To obtain (11.2), set $x = b/a$ in $(1+x)/2 \geq \sqrt{x}$ and multiply both sides by a.

Exercise 11.1. Use Figure 11.3 to give a purely geometric proof of (11.2). (Hint: compare the sum of the areas of the two isosceles right triangles to the area of the shaded rectangle.)

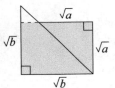

Figure 11.3. A geometric proof of (11.2)

Exercise 11.2. For a purely algebraic proof of (11.2), expand and simplify the inequality $(\sqrt{a} - \sqrt{b})^2 \geq 0$ (which is obviously true since squares are never negative).

Exercise 11.3. Show that (11.1) and (11.2) are equivalent. (Hint: let $x = \sqrt{a}/\sqrt{b}$ in (11.1) and clear fractions to obtain (11.2); then let $\{a, b\} = \{x, 1/x\}$ in (11.2) to obtain (11.1).)

Exercise 11.4. Find the minimum value of

$$\frac{(x+1/x)^6 - (x^6 + 1/x^6) - 2}{(x+1/x)^3 + (x^3 + 1/x^3)}$$

for $x > 0$. (Hint: simplify the expression by setting $a = (x+1/x)^3$ and $b = x^3 + 1/x^3$. This is problem B1 from the 1998 edition of the Putnam Mathematical Competition, which we discuss in Cameo 20.)

See Cameo 15 for some applications of the AM-GM inequality (11.2) to optimizations problems in calculus.

CAMEO 11. Tangent line inequalities

Exercise 11.5. Prove *Bernoulli's inequality* (named for Jacob Bernoulli, 1654–1705): Let $x > -1$. If $r > 1$ or $r < 0$, then
$$(1+x)^r \geq 1 + rx, \tag{11.3}$$
and if $0 < r < 1$, then
$$(1+x)^r \leq 1 + rx. \tag{11.4}$$
See Figure 11.4 for illustrations of Bernoulli's inequality.

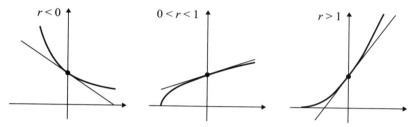

Figure 11.4. The three cases of Bernoulli's inequality

Exercise 11.6. Prove the *weighted AM-GM inequality*: Let $0 < r < 1$. If a and b are positive, then
$$\boxed{a^r b^{1-r} \leq ra + (1-r)b}$$
with equality if and only if $a = b$. (Hint: let $x = (a/b) - 1$ in (11.4) and multiply by b.) When $r = 1/2$ this inequality coincides with (11.2). See Cameo 36 for an application of the weighted AM-GM inequality to calculus.

The next three exercises in this Cameo lead to an expression for the natural logarithm as a limit:
$$\boxed{\text{if } x > 0 \text{ then } \ln x = \lim_{n \to \infty} n(x^{1/n} - 1).} \tag{11.5}$$

Exercise 11.7. Show that if $x > 0$ then $\ln x \leq x - 1 \leq x \ln x$. (Hint: see Figure 11.5.)

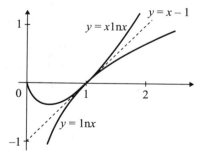

Figure 11.5. Graphs of $y = \ln x$, $y = x - 1$, and $y = x \ln x$

Exercise 11.8. Let $n > 0$. Show that
$$\ln x \leq n(x^{1/n} - 1) \leq x^{1/n} \ln x.$$

(Hint: multiply each term in the double inequality in Exercise 11.7 by n and then replace x with $x^{1/n}$.)

Exercise 11.9. Take the limit as $n \to \infty$ of each term in the double inequality in Exercise 11.8 and apply the squeeze theorem to obtain (11.5).

Exercise 11.10. (i) Show that $x \cos x < \sin x < x$ for x in $(0, \pi/2)$. (Hint: first show that $\sin x < x < \tan x$ for x in $(0, \pi/2)$. See Figure 11.6.)

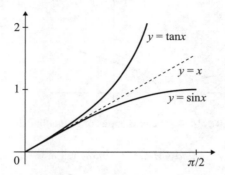

Figure 11.6. Graphs of $y = \sin x$, $y = x$, and $y = \tan x$

(ii) Prove *Aristarchus's inequalities*:

$$\text{if } 0 < \beta < \alpha < \pi/2, \text{ then } \frac{\sin \alpha}{\sin \beta} < \frac{\alpha}{\beta} < \frac{\tan \alpha}{\tan \beta}.$$

(Hint: it suffices to show that $(\sin x)/x$ is decreasing and $(\tan x)/x$ is increasing on $(0, \pi/2)$. Use (i) to determine the signs of the derivatives of $(\sin x)/x$ and $(\tan x)/x$.) The inequalities are named for the Greek mathematician and astronomer Aristarchus of Samos (circa 310–230 BCE).

Aristarchus of Samos

Jacob Bernoulli

CAMEO 12

A geometric illustration of the limit for e

Calculus texts define e in various ways: the number satisfying $\lim_{h \to 0} (e^h - 1)/h = 1$ or $\ln e = 1$ being common. But nearly every one goes on to show that

$$\lim_{n \to \infty} \left(1 + \frac{1}{n}\right)^n = e, \qquad (12.1)$$

often by an analytical argument, such as finding the derivative of the natural logarithm. But the limit (in the stronger form $\lim_{t \to 0} (1 + t)^{1/t} = e$) can be illustrated geometrically, as we now show.

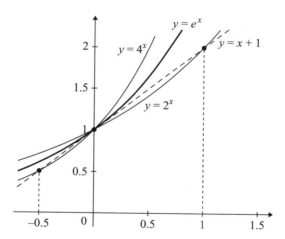

Figure 12.1. Graphs of three exponential functions

In Figure 12.1 we have graphed $y = b^x$ for three values of b: e, one number (2) smaller than e, one number (4) larger than e, and the line $y = x + 1$. The three curves and the line all pass through (0,1) where the line is tangent to $y = e^x$, but not tangent to $y = 2^x$ or to $y = 4^x$. For these two curves, $y = x + 1$ is a secant line and has a second point of intersection with the curve at a point (t, b^t) where

$$b^t = 1 + t. \qquad (12.2)$$

For example, when $b = 2$, $t = 1$ since $2^1 = 1 + 1$, and when $b = 4$, $t = -1/2$ since $4^{-1/2} = 1 + (-1/2)$.

27

When $t \neq 0$ we can solve (12.2) for b as a function of t:

$$b = (1+t)^{1/t}. \tag{12.3}$$

Moving t closer and closer to 0 changes the value of b for which the curve $y = b^x$ has a second intercept with $y = x+1$ until we reach (in the limit) the value of b for which $y = x+1$ is a tangent line, namely $b = e$. Hence $\lim_{t \to 0}(1+t)^{1/t} = e$.

When $t = 1/n$ for a positive integer n, we obtain (12.1). However, this geometric illustration of the limit has several other consequences. For example, we have the inequality $(1 + (1/n))^n < e$ since the values of b are smaller than e when t is positive in (12.3).

When $t = -1/(n+1)$ for a positive integer n, we obtain (after some algebra) the inequality $e < (1 + (1/n))^{n+1}$ since the values of b are larger than e when t is negative in (12.3). Combining the two inequalities yields for n positive

$$\left(1 + \frac{1}{n}\right)^n < e < \left(1 + \frac{1}{n}\right)^{n+1}. \tag{12.4}$$

With a geometric argument using the midpoint and trapezoidal rule approximations to a definite integral (see Cameo 29), the double inequality in (12.4) can be strengthened to

$$\left(1 + \frac{1}{n}\right)^{\sqrt{n(n+1)}} < e < \left(1 + \frac{1}{n}\right)^{n+\frac{1}{2}}$$

(the new exponents are the geometric and arithmetic means of n and $n + 1$).

SOURCE: Adapted from Karl Menger's 1955 text *Calculus: A Modern Approach*, recently reprinted by Dover.

CAMEO 13

Which is larger, e^π or π^e? a^b or b^a?

The first question in the title of this Cameo is easy to answer: get out your calculator, compute $e^\pi \approx 23.14$ and $\pi^e \approx 22.46$, and conclude that $e^\pi > \pi^e$. But while technology can answer the question, it doesn't shed much light on *why* the inequality $e^\pi > \pi^e$ is true.

In 1849 the Swiss mathematician Jakob Steiner (1796–1863) published the following problem in the *Journal für die reine und angewandte Mathematik* (the Journal for Pure and Applied Mathematics), better known as *Crelle's Journal* after its editor August Leopold Crelle (1780–1855): *For what positive value of x is the xth root of x the greatest?* Note that "the xth root of x" is related to $e^\pi > \pi^e$ since the inequality is equivalent to $e^{1/e} > \pi^{1/\pi}$. See Figure 13.1.

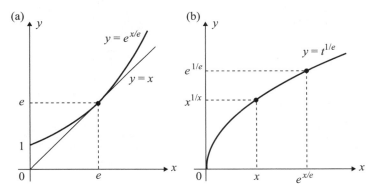

Figure 13.1. An illustration of the maximum value of $x^{1/x}$

In Figure 13.1a, we see that $e^{x/e} \geq x$ since the graph of $y = e^{x/e}$ is concave up and $y = x$ is its tangent line at (e, e). Taking the xth root of x and $e^{x/e}$ yields $e^{1/e} \geq x^{1/x}$ with equality if and only if $x = e$, as illustrated in Figure 13.1b for $x > 1$ (the other case differs only in concavity). Thus $e^{1/e} > \pi^{1/\pi}$ and consequently $e^\pi > \pi^e$.

To answer the question about a^b versus b^a (where $0 < a < b$), we need to know a little more about $y = x^{1/x}$. Using logarithmic differentiation, its derivative is $y' = x^{(1/x)}(1 - \ln x)/x^2$, so the graph is increasing for x in $(0,e)$ and decreasing for $x > e$. See Figure 13.2 for a graph of $y = x^{1/x}$ and its horizontal asymptote $y = 1$.

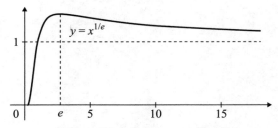

Figure 13.2. A graph of $y = x^{1/x}$

The graph yields the following conclusions (the cases where a or b equals 1 are trivial and omitted):

Case 1. For $e \leq a < b$, $a^{1/a} > b^{1/b}$ so that $a^b > b^a$. So for example, $2015^{2016} > 2016^{2015}$.

Case 2. For $0 < a < b \leq e$, $a^{1/a} < b^{1/b}$ so that $a^b < b^a$. So for example, $2.015^{2.016} < 2.016^{2.015}$.

Case 3. For $0 < a < 1 < b$, $a^b < b^a$ since $a^b < 1$ and $1 < b^a$.

Case 4. For $1 < a < e < b$, no general conclusion can be drawn, since $2^3 < 3^2$, $2^4 = 4^2$, and $2^5 > 5^2$.

Jakob Steiner

August Leopold Crelle

SOURCES: I. Niven, "Which is larger, e^π or π^e?" *Two-Year College Mathematics Journal*, **3** (1972), pp. 13–15; and RBN, "Proof without words: Steiner's problem on the number e," *Mathematics Magazine*, **82** (2009), p. 102.

CAMEO 14

Derivatives of area and volume

Many students notice that the derivative of the area A of a circle with respect to its radius r equals its circumference C, and that the derivative of the volume V of a sphere with respect to its radius r equals its surface area S:

$$\frac{dA}{dr} = \frac{d}{dr}\pi r^2 = 2\pi r = C \quad \text{and} \quad \frac{dV}{dr} = \frac{d}{dr}\frac{4}{3}\pi r^3 = 4\pi r^2 = S.$$

Consequently, students may ask: Are there other two-dimensional figures and three-dimensional solids with these properties?

Example 14.1. (a) An *annulus* is the region between two concentric circles. If the outer radius is twice the inner radius r, as shown in Figure 14.1a, then its area $A = \pi(2r)^2 - \pi r^2 = 3\pi r^2$ and its perimeter $P = 4\pi r + 2\pi r = 6\pi r$. Hence $dA/dr = P$. (b) In Figure 14.1b we see a circular cylinder whose height h equals the diameter of the base, so that its volume $V = \pi r^2(2r) = 2\pi r^3$ and its total surface area (the sum of the areas of the top, bottom, and the lateral or curved portion) is $S = 2(\pi r^2) + 2\pi r \cdot 2r = 6\pi r^2$. Hence $dV/dr = S$.

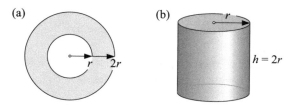

Figure 14.1. The annulus and cylinder in Example 14.1

Are the circle, sphere, annulus, and cylinder special in this regard, or do these relationships hold for other plane figures and solids? In this Cameo we learn the answer to this question, showing that it depends on how you choose to measure the linear dimensions of the objects.

Exercise 14.1. Consider a square and a cube each with side length s. Does the derivative of the area A of the square with respect to s equal the perimeter P? Is the derivative of the volume V of the cube with respect to s equal to the surface area S?

You discovered that the answer to each question is no, so perhaps the circle, sphere, annulus, and cylinder are special. But let's repeat the calculations using a different linear measurement.

Exercise 14.2. Repeat Exercise 14.1, but take the derivatives with respect to *half* the side length, i.e., set $s = 2x$ and find dA/dx and dV/dx.

31

Aha! Now $dA/dx = 8x = P$ and $dV/dx = 24x^2 = S$. So perhaps the "area-perimeter" and "volume-surface area" differentiation relationships do hold for objects other than the circle, sphere, annulus, and cylinder.

Consider a region in the plane whose perimeter P and area A are defined and finite, and expressible in terms of some linear dimension s of the region (such as side length, diameter, or radius) as $P = ks$ and $A = cs^2$ for constants k and c. We now find a linear dimension x (as a constant times s) such that $dA/dx = P$.

Let $x = ts$ for some constant t. Then $A = c(x/t)^2$ and $P = kx/t$, so that $dA/dx = P$ if and only if $t = 2c/k$, or equivalently, $x = 2cs/k = 2cs^2/ks = 2A/P$.

Exercise 14.3. Suppose the region in the plane is an equilateral triangle with side length s. Since its perimeter $P = 3s$ and area $A = (\sqrt{3}/4)s^2$, show that if $dA/dx = P$ then (a) $x = h/3$, where $h = s\sqrt{3}/2$ is the altitude of the triangle (geometrically, x is the radius of the inscribed circle), and (b) express P and A in terms of x and verify that $dA/dx = P$. See Figure 14.2a.

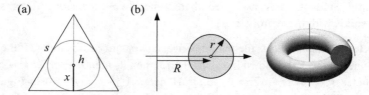

Figure 14.2. The equilateral triangle and torus in Exercises 14.3 and 14.5

Example 14.2. The results in Exercises 14.2 and 14.3 hold for regular n-gons. Let r denote the *inradius* (the radius of an inscribed circle) of the n-gon. Then the n-gon has side length $s = 2r\tan(\pi/n)$, perimeter $P = 2nr\tan(\pi/n)$, and area $A = nr^2\tan(\pi/n)$, hence $dA/dr = P$.

Exercise 14.4. Now consider a region in space whose surface area S and volume V are defined, finite, and expressible in terms of a linear dimension s as $S = ks^2$ and $V = cs^3$ for constants k and c. Show that in this case $dV/dx = S$ if and only if $x = 3V/S$.

Let's apply this result to the *torus* (perhaps from the Latin word for "knot"), a doughnut-shaped object (*doughnut* was originally spelled *dough knot*) obtained by revolving a circle about a line outside the circle. Let r denote the radius of the circle and R ($R > r$) the distance from the center of the circle to the axis of revolution, as shown in Figure 14.2b. Using integration, the volume V and the surface area S of the torus are $V = 2\pi^2 r^2 R$ and $S = 4\pi^2 rR$.

Exercise 14.5. Consider a torus for which $R = mr$ where $m > 1$ is a constant. (a) Show that $dV/dr \neq S$. (b) For what linear dimension x is $dV/dx = S$?

SOURCE: Adapted from J. Tong, "Area and perimeter, volume and surface area," *College Mathematics Journal*, **28** (1997), p. 57.

CAMEO 15

Means and optimization

Some of the most important applications of derivatives in the calculus course are the optimization problems. Many of these are geometric in nature, for example, finding the dimensions of a region in the plane with fixed perimeter that has maximum area, or finding the dimensions of an object with a fixed volume that has minimum surface area. However, for some of these problems there is a non-calculus method that is often simpler, and can be used to solve problems that can't be solved with single variable calculus.

The non-calculus method is based on the arithmetic mean-geometric mean (or AM-GM) inequality that we derived in Cameo 11. If a and b are positive numbers, then the arithmetic mean of a and b is $(a+b)/2$, the geometric mean is $\sqrt{ab}$, and the AM-GM inequality states that

$$\frac{a+b}{2} \geq \sqrt{ab} \tag{15.1}$$

with equality if and only if $a = b$. In Exercise 11.1 we presented a visual proof, and in Exercise 11.2 a simple algebraic proof. Let's see how this simple inequality (and its extension to three numbers) can be used to solve some optimization problems often encountered in a calculus course.

Example 15.1. Of all rectangles with a given perimeter, which one has the largest area? Of all rectangles with a given area, which one has the smallest perimeter? We can solve both parts of the problem by using the AM-GM inequality to construct an inequality between the perimeter P and the area A of a rectangle. If the lengths of the sides of the rectangle are x and y, then $P = 2(x+y)$, $A = xy$, and hence

$$\sqrt{A} = \sqrt{xy} \leq \frac{x+y}{2} = \frac{P}{4},$$

so that $A \leq (P/4)^2$, with equality if and only if $x = y$. So if the perimeter of the rectangle is given, then the square has maximum area, and if the area of the rectangle is given, then the square has minimum perimeter.

Exercise 15.1. Solve both parts of the problem in Example 15.1 using calculus.

Example 15.2. In 1471 Johannes Müller (1436–1476), called "Regiomontanus" after his birthplace Königsberg, wrote a letter to Christian Roder containing the following problem:

> At what point on the earth's surface does a perpendicularly suspended rod appear longest? (That is, at what point is the visual angle a maximum?)

In his classic work *100 Great Problems of Elementary Mathematics*, Heinrich Dörrie writes that this problem, now known as *Regiomontanus's maximum problem*, "deserves special attention as the *first extreme problem* encountered in the history of mathematics since the days of antiquity."

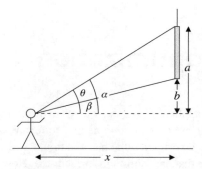

Figure 15.1. The Regiomontanus maximum problem

In Figure 15.1 we see the suspended rod, whose top and bottom are a and b units, respectively, above the eye level of the observer, x units away. The task is to find x to maximize the angle θ. Let α and β denote the angles that the lines of sight to the top and bottom of the rod make, respectively, with the observer's eye level. Then

$$\cot\theta = \frac{1}{\tan(\alpha-\beta)} = \frac{1+\tan\alpha\tan\beta}{\tan\alpha-\tan\beta}$$

$$= \frac{1+(a/x)(b/x)}{a/x - b/x} = \frac{x}{a-b} + \frac{ab}{(a-b)x}.$$

Since the cotangent is a decreasing function for θ in the first quadrant, to maximize θ we minimize $\cot\theta$. The AM-GM inequality now yields

$$\cot\theta = \frac{x}{a-b} + \frac{ab}{(a-b)x} \geq 2\sqrt{\frac{x}{a-b}\cdot\frac{ab}{(a-b)x}} = \frac{2\sqrt{ab}}{a-b},$$

with equality if and only if $x/(a-b) = ab/(a-b)x$, or $x = \sqrt{ab}$. Thus the observer should stand at a distance equal to the geometric mean of the heights of the top and bottom of the rod above the observer's eye level.

Exercise 15.2. Use calculus to solve Regiomontanus's maximum problem in Example 15.2.

Many optimization problems involve volumes of objects. Since volume is a three dimensional concept, we may be able to use the AM-GM inequality for three numbers: If a, b, and c are positive numbers, then the arithmetic mean is $(a+b+c)/3$, the geometric mean is $\sqrt[3]{abc}$, and

$$\boxed{\sqrt[3]{abc} \leq \frac{a+b+c}{3}} \tag{15.2}$$

with equality if and only if $a = b = c$. See Figure 15.2 for a visual proof.

CAMEO 15. Means and optimization

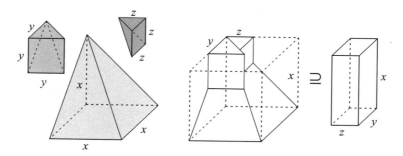

Figure 15.2. The AM-GM inequality for three numbers

In Figure 15.2 (a three-dimensional version of Figure 11.3) we see how a box with sides x, y, and z fits inside the union of three right pyramids whose bases are squares with sides x, y, and z and whose altitudes are also x, y, and z, respectively. Since the volume of a pyramid is $1/3$ the area of the base times the height, we have

$$xyz \leq \frac{1}{3}x^2 \cdot x + \frac{1}{3}y^2 \cdot y + \frac{1}{3}z^2 \cdot z = \frac{x^3 + y^3 + z^3}{3}.$$

Setting $x = \sqrt[3]{a}$, $y = \sqrt[3]{b}$, and $z = \sqrt[3]{c}$ yields (15.2).

Exercise 15.3. Prove (15.2) algebraically. (Hint: first establish the identity

$$x^3 + y^3 + z^3 - 3xyz = (x + y + z)(x^2 + y^2 + z^2 - xy - yz - xz),$$

then use the AM-GM inequality (15.1) for two numbers to show that the second term on the right is nonnegative for positive x, y, and z, and set $x = \sqrt[3]{a}$, $y = \sqrt[3]{b}$, and $z = \sqrt[3]{c}$ as before.)

The AM-GM inequality (15.2) for three numbers is a powerful problem-solving tool. We illustrate its use with four examples—the first and fourth can be solved with single-variable calculus, but the second and third require multivariable calculus.

Example 15.3. Years ago tents were made of canvas and shaped like cones, as seen in the photograph in Figure 15.3 of the Native American encampment at the Pan-American Exposition in Buffalo, NY in 1901.

Among all possible conical canvas tents with a specified volume and no floor, what is the ratio of the height h to the base radius r to minimize the amount of canvas used?

The volume of the conical tent is $V = \pi r^2 h/3$ and the amount S of canvas is the lateral area $S = \pi r \sqrt{r^2 + h^2}$ of the cone. To minimize S it suffices to minimize S^2. To that end, we have

$$S^2 = \pi^2 r^2 (r^2 + h^2) = \pi^2 (r^4 + r^2 h^2)$$

$$= \pi^2 \left(r^4 + \frac{r^2 h^2}{2} + \frac{r^2 h^2}{2} \right) \geq 3\pi^2 \left(r^4 \cdot \frac{r^2 h^2}{2} \cdot \frac{r^2 h^2}{2} \right)^{1/3} = 3\pi^2 \left(\frac{3V}{\pi \sqrt{2}} \right)^{4/3}$$

with equality if $r^4 = r^2 h^2/2$, or equivalently, $h = r\sqrt{2}$. Thus the ratio of h to r should be $\sqrt{2}$ to minimize the amount of canvas used to make the tent.

Figure 15.3. Conical tents

Exercise 15.4. Use calculus to solve the problem in Example 15.3.

In the above problem we *minimized a sum* by expressing the sum as a *sum of terms with a constant product*. In the next problem we *maximize a product* by expressing the product as a *product of terms with a constant sum*.

Example 15.4. A well-known package delivery service restricts the size of packages it will accept. Packages cannot exceed 108 inches in length plus girth, i.e., length + 2 · width + 2 · height ≤ 108. Find the dimensions of an acceptable package with maximum volume. See Figure 15.4.

Figure 15.4. Designing an acceptable package with maximum volume

Let x = length, y = width, and z = height of the rectangular box in Figure 15.4. Then the sum of the length and girth is $S = x + 2y + 2z$. If V denotes the volume, then $4V = 4xyz = x \cdot 2y \cdot 2z$. From (15.2) we have

$$\sqrt[3]{4V} = \sqrt[3]{x \cdot 2y \cdot 2z} \leq \frac{x + 2y + 2z}{3} = \frac{S}{3} \leq \frac{108}{3} = 36$$

CAMEO 15. Means and optimization

and so the volume satisfies $V \leq 36^3/4 = 11664$ in^3 with equality if $x = 2y = 2z$. Hence the design of the acceptable rectangular box with maximum volume is given by $x^3/4 = 11664$, so that $x = 36$ in, $y = z = 18$ in. (To solve this problem with single-variable calculus requires another assumption, such as a square end for the package.)

Note that a cubical box 22 in on a side with volume 10648 in^3 is unacceptable (since its length plus girth is 110 in) although the volume of the cube is over 1000 in^3 less than the volume of the acceptable package described above.

Example 15.5. Plastic boxes with compartments, such as the one with 24 compartments in Figure 15.5, are common for storing small items in the home, office, or workplace.

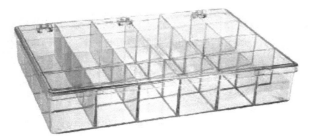

Figure 15.5. A plastic box with 24 compartments

How should such a box (with 24 compartments and a lid) be designed (length, width, and height) if it is to have a volume of 560 in^3 and use as little plastic as possible?

Let x = length, y = width, and z = height in inches, with volume $V = 560$ in^3. The amount P of plastic used is proportional to the total area of the top, bottom, sides, and partitions in the box, so $P = c(2xy + 5xz + 7yz)$ for some constant c. Hence

$$(70V^2)^{1/3} = \sqrt[3]{2xy \cdot 5xz \cdot 7yz} \leq \frac{2xy + 5xz + 7yz}{3} = \frac{P}{3c}$$

with equality if and only if $2xy = 5xz = 7yz$, or $x = 7z/2$ and $y = 5z/2$. Hence $V = 560 = \frac{7}{2}z \cdot \frac{5}{2}z \cdot z = \frac{35}{4}z^3$ so that $z = 4$ in, $x = 14$ in, and $y = 10$ in. However, in such a box, the dimensions of the compartments are $2\frac{1}{3}$ in by $2\frac{1}{2}$ in by 4 in, so boxes like the one in Figure 15.6 are most likely not designed to minimize the amount of plastic used for their construction. (As in Example 15.4, to solve this problem with single-variable calculus one usually has to make an additional assumption, such as the box or its compartments have square bases.)

Example 15.6. Supermarkets sell ground coffee in cylindrical cans constructed of three different materials, a metal bottom, cardboard sides, and a plastic cap, as shown in Figure 15.6. If the costs of metal, cardboard, and plastic are m, c, and p cents per in^2 respectively, what should the ratio of the height h to the base radius r be in order to minimize the construction cost for a can with volume V?

The volume V and cost C are given by $V = \pi r^2 h$ in^3 and $C = (p+m)\pi r^2 + 2c\pi rh$ cents. Hence

$$\sqrt[3]{c^2\pi(p+m)V^2} = \sqrt[3]{(p+m)\pi r^2 \cdot c\pi rh \cdot c\pi rh}$$
$$\leq \frac{(p+m)\pi r^2 + c\pi rh + c\pi rh}{3} = \frac{C}{3},$$

Figure 15.6. A coffee can

with equality if and only if $(p+m)\pi r^2 = c\pi rh$, or $h/r = (p+m)/c$. Since it is reasonable to assume that both p and m are greater than c, it follows that these cans should have $h/r > 2$, which often appears to be the case.

Exercise 15.5. Use calculus to solve the problem in Example 15.6.

Exercise 15.6. In Example 15.1 we derived an inequality between the area and the perimeter of a rectangle. Now consider a rectangular box with dimensions a, b, and c. Let $V = abc$ be its volume, $S = 2(ab + bc + ac)$ the total area of the six faces, and $E = 4(a + b + c)$ the total length of the twelve edges. Prove that

$$6V^{2/3} \leq S \leq E^2/24,$$

with equality if and only if the box is a cube.

Exercise 15.7. Prove the AM-GM inequality for four positive numbers $a, b, c,$ and d:

$$\boxed{\sqrt[4]{abcd} \leq \frac{a+b+c+d}{4}} \qquad (15.3)$$

with equality if and only if $a = b = c = d$. (Hint: use (15.1) twice.)

The AM-GM inequalities in (15.1), (15.2), and (15.3) extend to n positive numbers. See Cameo 45 for a calculus-based proof.

PART II
Integration

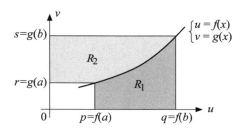

CAMEO 16

Combinatorial identities for Riemann sums

Almost all current calculus texts, when introducing the definite integral, ask students to evaluate a few by hand, usually for polynomial integrands. To do so, the students need several combinatorial identities, including these three:

$$\sum_{i=1}^{n} i = 1 + 2 + 3 + \cdots + n = \frac{n(n+1)}{2} \tag{16.1}$$

$$\sum_{i=1}^{n} i^2 = 1^2 + 2^2 + 3^2 + \cdots + n^2 = \frac{n(n+1)(2n+1)}{6} \tag{16.2}$$

$$\sum_{i=1}^{n} i^3 = 1^3 + 2^3 + 3^3 + \cdots + n^3 = \left[\frac{n(n+1)}{2}\right]^2. \tag{16.3}$$

When proofs are presented they are often by mathematical induction. However, while induction will verify a correct formula, students may tell you that induction doesn't say much about *why* the formula is true. As the Italian-American mathematician Gian-Carlo Rota (1932–1999) put it, "If we have no idea why a statement is true, we can still prove it by induction."

Whether or not students see a proof of the above formulas, they should, at a minimum, be convinced of the truth of each one. And if seeing is believing, perhaps a visual argument or two for the truth of each one is in order.

In the next three examples we illustrate (16.1), (16.2), and (16.3).

Example 16.1. In Figure 16.1a we represent the sum $1 + 2 + 3 + \cdots + n$ with a collection of balls in a triangular array with n rows.

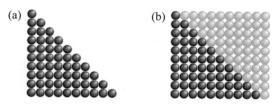

Figure 16.1. Representing $1 + 2 + \cdots + n = n(n+1)/2$

In Figure 16.1b we see how two such triangular arrays, each with $1 + 2 + 3 + \cdots + n$ balls, form a rectangular array with n rows and $n + 1$ columns, so that $2(1 + 2 + 3 + \cdots + n) = n(n + 1)$, from which (16.1) follows. The sum $1 + 2 + 3 + \cdots + n = n(n + 1)/2$ in (16.1) is often called a *triangular number*, as illustrated in Figure 16.1a.

Carl Friedrich Gauss and the 100th triangular number

Nearly every biography of the great mathematician Carl Friedrich Gauss (1777–1855) relates the following story. When Gauss was about ten years old, his arithmetic teacher asked the students in class to compute the sum $1 + 2 + 3 + \cdots + 100$, anticipating this would keep them busy for some time. He barely finished stating the problem when young Carl came forward and placed his slate on the teacher's desk, void of calculation, with the correct answer: 5050. When asked to explain, Gauss admitted he recognized the pattern $1 + 100 = 101, 2 + 99 = 101, 3 + 98 = 101$, and so on to $50 + 51 = 101$. Since there are fifty such pairs, the sum must be $50 \cdot 101 = 5050$. The pattern for the sum (adding the largest number to the smallest, the second largest to the second smallest, and so on) is illustrated below, along with a portrait of Gauss on a pre-euro 10 Deutsche Mark note.

Gauss and his computation

For a second illustration of (16.1) we represent the sum $1 + 2 + 3 + \cdots + n$ as the area of a triangular array of unit squares (squares with area 1), as shown in Figure 16.2a.

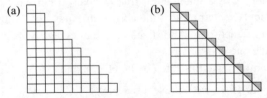

Figure 16.2. A second illustration that $1 + 2 + \cdots + n = n(n + 1)/2$

In Figure 16.2b we compute the total area of the array a second way, as the area $\frac{1}{2}n \cdot n = n^2/2$ of the large white triangle plus the total area $n \cdot \frac{1}{2} = n/2$ of the n small gray triangles each with area $1/2$, so that

$$1 + 2 + 3 + \cdots + n = \frac{n^2}{2} + \frac{n}{2} = \frac{n(n+1)}{2}.$$

CAMEO 16. Combinatorial identities for Riemann sums

Example 16.2. To represent a sum of squares, we consider a collection of n squares with areas $1^2, 2^2, \ldots, n^2$. Three such collections are shown (for $n = 5$) in Figure 16.3a.

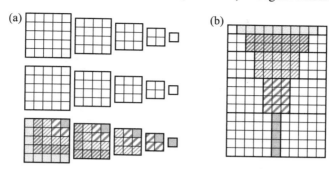

Figure 16.3. Illustrating the sum of squares formula

If we slice up the third collection of squares as indicated by the different shadings, all the squares can be rearranged into the rectangle in Figure 16.3b with height $1 + 2 + 3 + \cdots + n = n(n+1)/2$ and base $n + 1 + n = 2n + 1$, so that

$$3(1^2 + 2^2 + 3^2 + \cdots + n^2) = \frac{n(n+1)}{2} \cdot (2n+1) = \frac{n(n+1)(2n+1)}{2},$$

from which (16.2) follows upon division by 3.

Our second illustration of (16.2) uses three stacks of n layers of unit cubes (cubes with volume 1) where the volumes of the layers are $1^2, 2^2, \ldots, n^2$, as shown (for $n = 4$) in Figure 16.4a. Each stack of cubes has a pyramidal shape so that the three stacks fit together as shown in Figure 16.4b.

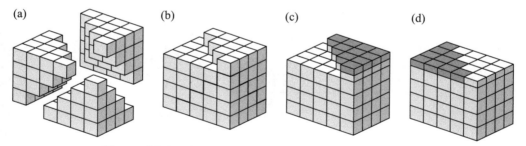

Figure 16.4. A second illustration of the sum of squares formula

In Figure 16.4c we slice the top layer of cubes horizontally, and place the darker gray half-cubes on top of the unshaded portion of the stack, to form a rectangular box of cubes with base n by $n + 1$ and height $n + (1/2)$ in Figure 16.4d, so that

$$3(1^2 + 2^2 + 3^2 + \cdots + n^2) = n(n+1)\left(n + \frac{1}{2}\right) = \frac{n(n+1)(2n+1)}{2},$$

and again (16.2) follows after division by 3. If the three-dimensional argument is difficult to visualize, you may want to illustrate Figure 16.4 with a collection of small plastic cubes, which can be obtained from educational supply houses.

Example 16.3. Here we represent k^3 by a $k \times k \times k$ collection of unit cubes. Slice n such cubes (with volumes $1^3, 2^3, \ldots, n^3$) into layers and arrange on a plane, as shown (for $n = 4$) in Figure 16.5a.

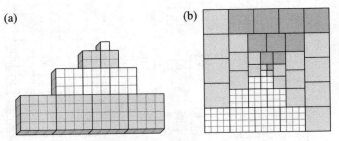

Figure 16.5. Illustrating sums of cubes

Now arrange four copies of the cubes in Figure 16.5a into a square pattern with side length $n \cdot n + n = n(n + 1)$ as shown in Figure 16.5b, so that

$$4(1^3 + 2^3 + 3^3 + \cdots + n^3) = [n(n + 1)]^2$$

from which (16.3) follows. Combining (16.3) and (16.1) yields this attractive formula relating the sum of cubes to the squared sum of the first n positive integers:

$$1^3 + 2^3 + 3^3 + \cdots + n^3 = (1 + 2 + 3 + \cdots + n)^2. \tag{16.4}$$

In Figure 16.6 we see an illustration of (16.4) representing k^3 as k copies of k^2 for k from 1 to n (shown here for $n = 6$). When k is even two squares overlap, but the area of the overlap is the same as the area of a square (in white) not covered by the shaded squares.

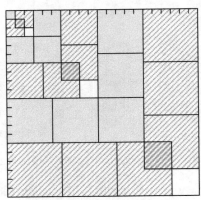

Figure 16.6. Illustrating (16.4)

SOURCES:

Figure 16.2: I. Richards, "Proof without words: Sum of integers'" *Mathematics Magazine*, **57** (1984), p. 104.

Figure 16.3: M. Gardner, "Mathematical games," *Scientific American*, **229** (1973), p. 115, and D. Kalman, "$(1 + 2 + \cdots + n)(2n + 1) = 3(1^2 + 2^2 + \cdots + n^2)$", *College Mathematics Journal*, **22** (1991), p. 124.

Figure 16.4: M.-K. Siu, "Proof without words: Sum of squares," *Mathematics Magazine*, **57** (1984), p. 92.

Figure 16.5: A. Cupillari, "Proof without words: $1^3 + 2^3 + \cdots + n^3 = (n(n+1))^2/4$" *Mathematics Magazine*, **62** (1989), p. 259, and W. Lushbaugh, *Mathematical Gazette*, **49** (1965), p. 200.

Figure 16.6: S. Golomb, "A geometric proof of a famous identity," *Mathematical Gazette*, **49** (1965), pp. 198–200.

CAMEO 17

Summation by parts

Summation by parts is a deceptively simple yet remarkably powerful method for computing certain sums in calculus, and can be used in higher level courses as well. For two sequences $a_1, a_2, a_3, \ldots$ and $b_0, b_1, b_2, \ldots$ of real numbers, we have

$$\sum_{i=1}^{n} b_i(a_{i+1} - a_i) + \sum_{i=1}^{n} a_i(b_i - b_{i-1}) = a_{n+1}b_n - a_1 b_0. \qquad (17.1)$$

Expanding the sums and observing that almost all of the terms in the first sum cancel with terms in the second sum readily verifies the formula. In Figure 17.1 we have a geometric illustration of (17.1) for the case when both sequences are increasing sequences of positive numbers, so that each summand represents the area of a rectangle.

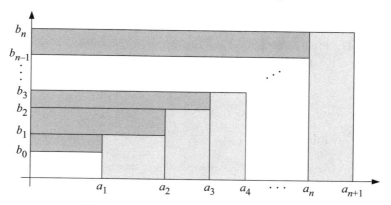

Figure 17.1. Illustrating summation by parts

A similar picture will be used to illustrate a continuous version of (17.1)—*integration by parts*—in the next Cameo. We now examine several applications. The first three are the sums of integers, squares, and cubes illustrated in the preceding Cameo.

Example 17.1. If we set $a_i = b_i = i$, then $a_{i+1} - a_i$ and $b_i - b_{i-1}$ are equal to 1 for every i, and (17.1) yields

$$\sum_{i=1}^{n} i + \sum_{i=1}^{n} i = (n+1)n$$

and hence $\sum_{i=1}^{n} i = \frac{n(n+1)}{2}$, as in (16.1) in the preceding Cameo.

Example 17.2. To verify the formula (16.2) for the sum of squares of the first n positive integers, we use the sequences of odd numbers and triangular numbers in (16.1), setting $a_i = 2i - 1$ and $b_i = i(i+1)/2$. Then $a_{i+1} - a_i = 2$ (the difference of consecutive odd numbers is 2) and $b_i - b_{i-1} = i$ (since $b_i = b_{i-1} + i$), and (17.1) yields

$$\sum_{i=1}^{n} \frac{i(i+1)}{2} \cdot 2 + \sum_{i=1}^{n} (2i-1) \cdot i = (2n+1) \cdot \frac{n(n+1)}{2} - 1 \cdot 0,$$

or

$$\sum_{i=1}^{n} (i^2 + i) + \sum_{i=1}^{n} (2i^2 - i) = \frac{n(n+1)(2n+1)}{2}.$$

Thus $3 \sum_{i=1}^{n} i^2 = \frac{n(n+1)(2n+1)}{2}$, so that $\sum_{i=1}^{n} i^2 = \frac{n(n+1)(2n+1)}{6}$.

Exercise 17.1. Show that $\sum_{i=1}^{n} (2i - 1) = n^2$, i.e., the sum of the first n odd numbers is n^2. See Figure 17.2. (Hint: try $a_i = i - 1$ and $b_i = i$.)

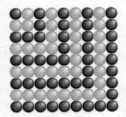

Figure 17.2. The sum of the first 9 odd numbers is 9^2

Example 17.3. For (16.3), the sum of the first n cubes, we set $a_i = b_i = i^2$. From the preceding exercise, the difference between consecutive squares is an odd number, and hence

$$\sum_{i=1}^{n} i^2(2i+1) + \sum_{i=1}^{n} i^2(2i-1) = (n+1)^2 \cdot n^2 - 1 \cdot 0$$

so that $4 \sum_{i=1}^{n} i^3 = [n(n+1)]^2$, or $\sum_{i=1}^{n} i^3 = \frac{[n(n+1)]^2}{4}$.

Exercise 17.2. Find the sum $\sum_{i=1}^{n} \frac{i(i+1)}{2}$ of the first n triangular numbers. (Hint: set $a_i = i+1$ and $b_i = \frac{i(i+1)}{2}$.) Although this one isn't needed in calculus, it's good practice using (17.1).

Exercise 17.3. Evaluate $\sum_{i=1}^{n} i^4$. (Hint: set $a_i = 3i(i-1)$ and $b_i = \frac{i(i+1)(2i+1)}{6}$.)

Summation by parts can also be used to find formulas for partial sums of certain infinite series, such as geometric series and telescoping series. We illustrate with the next two examples and the final two exercises.

Example 17.4. To use (17.1) to find the formula for the nth partial sum $\sum_{i=1}^{n} ar^{i-1}$ of the geometric series $\sum_{i=1}^{\infty} ar^{i-1}$, we set $a_i = a$ and $b_i = r^i$ ($r \neq 1$). Then (16.1) yields

$$\sum_{i=1}^{n} a(r^i - r^{i-1}) = ar^n - a,$$

so that $(r-1) \sum_{i=1}^{n} ar^{i-1} = a(r^n - 1)$ and hence $\sum_{i=1}^{n} ar^{i-1} = a(1 - r^n)/(1 - r)$.

CAMEO 17. Summation by parts

Example 17.5. We can use (17.1) to find the nth partial sum of a telescoping series without having to use partial fractions to simplify the summand. To find the nth partial sum $\sum_{i=1}^{n} \frac{1}{i(i+1)}$ of the series $\sum_{i=1}^{\infty} \frac{1}{i(i+1)}$, set $a_i = \frac{-1}{i(i+1)}$ and $b_i = i+2$. Then $a_{i+1} - a_i = \frac{2}{i(i+1)(i+2)}$ and $b_i - b_{i-1} = 1$ so that

$$\sum_{i=1}^{n}(i+2) \cdot \frac{2}{i(i+1)(i+2)} + \sum_{i=1}^{n} \frac{-1}{i(i+1)} \cdot 1 = \frac{-1}{(n+1)(n+2)} \cdot (n+2) - \frac{-1}{2} \cdot 2,$$

and hence $\sum_{i=1}^{n} \frac{1}{i(i+1)} = 1 - \frac{1}{n+1}$.

Exercise 17.4. Find a formula for the nth partial sum of the series $\sum_{i=1}^{\infty} \frac{1}{i(i+1)(i+2)}$. (Hint: set $a_k = \frac{-1}{i(i+1)(i+2)}$ and $b_i = i+3$.)

The final exercise shows that summation by parts can be used to find the formula for the partial sums of an infinite series that may be difficult to obtain by other methods.

Exercise 17.5. Show that the nth partial sum of the series $\sum_{i=1}^{\infty} i(1/2)^{i-1}$ is $4 - (n+2)(1/2)^{n-1}$. (Hint: set $a_i = (1/2)^{i-1}$, $b_i = i$, and use the result of Example 17.4.)

SOURCE: Adapted from G. Fredricks and RBN, "Summation by parts," *College Mathematics Journal*, **23** (1992), pp. 39–42.

CAMEO 18

Integration by parts

Let f and g be two functions with continuous derivatives. The definite integral version of the integration by parts formula

$$\int_a^b f(x)g'(x)dx = f(x)g(x)|_a^b - \int_a^b g(x)f'(x)dx \qquad (18.1)$$

can be illustrated geometrically in the special case when f and g are both positive and increasing on the interval $[a,b]$. Let $u = f(x)$ and $v = g(x)$, and sketch a graph of the curve (u,v) in the uv-plane, as shown in Figure 18.1. Since this is a curve given parametrically (a topic usually appearing later in a calculus course), students may wish to think of x as time, with both the u and v coordinates of the curve changing over the time interval $[a,b]$.

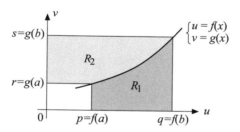

Figure 18.1. Integration by parts

The area of the dark gray region R_1 is $\int_p^q v\,du$ (the rectangles in the Riemann sum for this integral have, in the limit, height v and width du), and since $u = f(x)$ and $v = g(x)$, we have

$$\int_p^q v\,du = \int_a^b g(x)f'(x)dx.$$

Similarly the area of the light gray region R_2 is

$$\int_r^s u\,dv = \int_a^b f(x)g'(x)dx.$$

The sum of the areas of the two regions equals the difference of the areas of two rectangles, so that

$$\int_a^b f(x)g'(x)dx + \int_a^b g(x)f'(x)dx = f(b)g(b) - f(a)g(a) = f(x)g(x)|_a^b,$$

from which (18.1) follows.

Exercise 18.1. Establish (18.1) in the case where f is increasing and g decreasing, as shown in Figure 18.2.

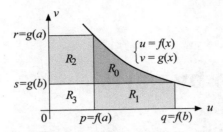

Figure 18.2. Integration by parts, a second case

(Hint: if A_k denotes the area of R_k, compute $A_1 - A_2$ in two ways: $(A_0 + A_1) - (A_0 + A_2)$ and $(A_1 + A_3) - (A_2 + A_3)$.)

SOURCE: R. Courant, *Differential and Integral Calculus*, Vol. 1, Interscience, New York, 1937.

CAMEO 19

The world's sneakiest substitution

In his book *Calculus* Michael Spivak writes "The world's sneakiest substitution is undoubtedly $z = \tan(x/2)$," used in calculus to integrate rational functions of the sine and cosine. It leads to the substitution

$$\sin x = \frac{2z}{1+z^2}, \quad \cos x = \frac{1-z^2}{1+z^2}, \quad \text{and} \quad dx = \frac{2dz}{1+z^2} \tag{19.1}$$

so that a rational function of $\sin x$ and $\cos x$ becomes a rational function of z that can often be integrated using the partial fractions technique. The substitution is known more formally as the *Weierstrass substitution*, named for the German mathematician Karl Theodor Wilhelm Weierstrass (1815–1897).

Most calculus texts derive (19.1) analytically, but perhaps a geometric illustration for small positive values of x can provide a motivation for the proof. See Figure 19.1.

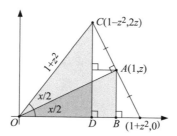

Figure 19.1. The world's sneakiest substitution

Notice that $z = \tan(x/2)$ (in $\triangle OAB$) implies that $\sin x = 2z/(1+z^2)$ and $\cos x = (1-z^2)/(1+z^2)$ (in $\triangle OCD$). To obtain the formula for dx, write $x = 2\arctan z$ and evaluate the differential of x.

While the picture in Figure 19.1 isn't a proof, it does suggest one: use $z = \tan(x/2)$ for x in $(-\pi, \pi)$ to evaluate $\sin(x/2)$ and $\cos(x/2)$, and then use double-angle formulas to find $\sin x$ and $\cos x$. In $\triangle OAB$ we have

$$\sin(x/2) = \frac{z}{\sqrt{1+z^2}} \quad \text{and} \quad \cos(x/2) = \frac{1}{\sqrt{1+z^2}},$$

so that the double angle formulas (see Cameo 49) yield

$$\sin x = 2\sin(x/2)\cos(x/2) = \frac{2z}{1+z^2} \quad \text{and} \quad \cos x = \cos^2(x/2) - \sin^2(x/2) = \frac{1-z^2}{1+z^2},$$

as seen in $\triangle OCD$ in Figure 19.1.

Exercise 19.1. Use Figure 19.2 to give another proof of (19.1). (Hint: express $|AC|$ in terms of z and note that the two shaded triangles are similar.)

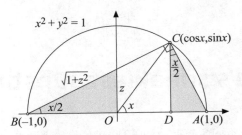

Figure 19.2. A second proof of (19.1)

Exercise 19.2. In Figure 19.3 we see a trapezoid consisting of three similar right triangles. Use the figure to give a third proof of (19.1). (Hint: express the lengths of the hypotenuses of the right triangles in terms of z.)

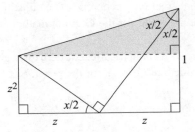

Figure 19.3. A third proof of (19.1)

Karl Theodor Wilhelm Weierstrass

SOURCES:

Figure 19.1 is adapted from RBN, "Proof without words: The substitution to make a rational function of the sine and cosine," *Mathematics Magazine*, **62** (1989), p. 267.

Figure 19.2 is from P. Deiermann, "The method of last resort (Weierstrass substitution)", *College Mathematics Journal*, **29** (1998), p. 17.

Figure 19.3 is adapted from S. H. Kung, "Proof without words: The Weierstrass substitution," *Mathematics Magazine*, **74** (2001), p. 393.

CAMEO 20

Symmetry and integration

Consider the integral

$$\int_0^{2\pi} \frac{1}{1+e^{\sin x}}\,dx. \tag{20.1}$$

None of the techniques in the traditional calculus course for finding antiderivatives seem to help here. But before resorting to a numeric method such as the midpoint, trapezoidal, or Simpson's rule, take a look at the graph of the integrand. Since it is positive on $[0, 2\pi]$, the integral represents the area of the shaded region in Figure 20.1.

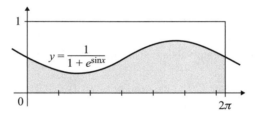

Figure 20.1. A graph of the region whose area is given by (20.1)

The function appears to have symmetry similar to the symmetry of an odd function, but the center of symmetry is not the origin but rather the point $(\pi, 1/2)$. Hence we suspect that the value of the integral in (20.1) should equal one-half the area of the enclosing rectangle with width 2π and height 1, that is, π.

If the graph of $y = f(x)$ on $[a,b]$ is symmetric with respect to the point $((a+b)/2, f((a+b)/2))$, then

$$\text{for all } x \text{ in } [a,b], \frac{1}{2}[f(x) + f(a+b-x)] = f\left(\frac{a+b}{2}\right). \tag{20.2}$$

We illustrate this symmetry condition in Figure 20.2.

57

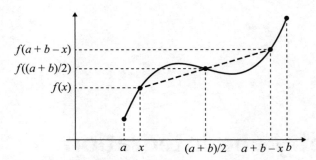

Figure 20.2. Illustrating the symmetry condition (20.2)

Then we have

$$\int_a^b f(x)dx = \frac{1}{2}\int_a^b f(x)dx + \frac{1}{2}\int_a^b f(t)dt = \frac{1}{2}\int_a^b f(x)dx - \frac{1}{2}\int_b^a f(a+b-x)dx$$

$$= \int_a^b \frac{1}{2}[f(x)dx + f(a+b-x)]dx = \int_a^b f\left(\frac{a+b}{2}\right)dx = (b-a)f\left(\frac{a+b}{2}\right).$$

So if (20.2) holds, then

$$\int_a^b f(x)dx = (b-a)f\left(\frac{a+b}{2}\right), \qquad (20.3)$$

that is, the midpoint rule for numeric integration with a single interval is exact.

Exercise 20.1. Show that when (20.2) holds, the trapezoidal rule with a single interval and Simpson's rule with two intervals are also exact for $\int_a^b f(x)dx$.

Example 20.1. So, to evaluate (20.1) we need only verify (20.2) for the integrand and employ (20.3). For $f(x) = 1/(1 + e^{\sin x})$ on $[0, 2\pi]$ we have

$$\frac{1}{2}\left[\frac{1}{1+e^{\sin x}} + \frac{1}{1+e^{\sin(2\pi-x)}}\right] = \frac{1}{2}\left[\frac{1}{1+e^{\sin x}} + \frac{1}{1+e^{-\sin x}}\right]$$

$$= \frac{1}{2}\left[\frac{1}{1+e^{\sin x}} + \frac{e^{\sin x}}{e^{\sin x}+1}\right] = \frac{1}{2} = \frac{1}{1+e^{\sin \pi}},$$

so (20.2) holds and hence $\int_0^{2\pi} \frac{1}{1+e^{\sin x}} dx = 2\pi \cdot \frac{1}{2} = \pi$.

Here are some to try (answers are in parentheses). We recommend using a graphing calculator first to observe the symmetry of the integrand in each, and then using (20.2) and (20.3) to evaluate the integral.

Exercise 20.2. Evaluate $\int_0^4 \frac{dx}{4+2^x}$ $(1/2)$.

Exercise 20.3. Evaluate $\int_{-1}^1 \arctan(e^x)\, dx$ $(\pi/2)$.

Exercise 20.4. Evaluate $\int_{-1}^1 \arccos(x^3)dx$ (π).

Exercise 20.5. Evaluate $\int_0^2 \frac{dx}{x+\sqrt{x^2-2x+2}}$ (1).

CAMEO 20. Symmetry and integration

The William Lowell Putnam Mathematical Competition

William Lowell Putnam (1861–1923), a member of the Harvard class of 1882, believed in the value of team competition in academics. In 1927, Putnam's widow Elizabeth established a trust fund to support the *William Lowell Putnam Mathematical Competition*, a challenging proof-oriented annual mathematics contest for college and university students in the United States and Canada. An article in *Time* magazine (December 16, 2002) called it the "world's toughest math test." Since 1962 "the Putnam," as it has become known, consists of twelve problems in two sessions of three hours each, held on the first Saturday in December.

William Lowell Putnam

Here are three problems from the Putnam that can be solved by exploiting the symmetry of the integrand.

Example 20.2. [Problem A5, 2005]. Evaluate $\int_0^1 \frac{\ln(x+1)}{x^2+1} dx$. While this integrand is not symmetric on [0,1], the presence of $x^2 + 1$ in the denominator suggests the substitution $\theta = \arctan x$, $\tan \theta = x$, and $d\theta = dx/(x^2 + 1)$ so that

$$\int_0^1 \frac{\ln(x+1)}{x^2+1} dx = \int_0^{\pi/4} \ln(\tan \theta + 1) d\theta.$$

Next we verify (20.2) for the integrand $f(\theta) = \ln(\tan \theta + 1)$ on $[0, \pi/4]$ and employ (20.3):

$$\frac{1}{2}\left[\ln(\tan \theta + 1) + \ln\left(\tan\left(\frac{\pi}{4} - \theta\right) + 1\right)\right] = \frac{1}{2}\left[\ln(\tan \theta + 1) + \ln\left(1 + \frac{1 - \tan x}{1 + \tan x}\right)\right]$$

$$= \frac{1}{2}\ln 2 = \ln\left(1 + \tan\frac{\pi}{8}\right)$$

since $\tan\frac{\pi}{8} = \sqrt{2} - 1$. Hence $\int_0^{\pi/4} \ln(\tan \theta + 1) d\theta = \frac{\pi}{4} \cdot \frac{1}{2} \ln 2 = \frac{\pi}{8} \ln 2$.

Exercise 20.6. [Problem A3, 1980]. Evaluate $\int_0^{\pi/2} \frac{1}{1+(\tan x)^{\sqrt{2}}} dx$.

Exercise 20.7. [Problem B1, 1987]. Evaluate $\int_2^4 \frac{\sqrt{\ln(9-x)}}{\sqrt{\ln(9-x)} + \sqrt{\ln(x+3)}} dx$.

By the way, calculators, graphing or otherwise, are not permitted in the Putnam Competition!

SOURCE: Adapted from RBN, "Symmetry and integration," *College Mathematics Journal*, **26** (1995), pp. 39–41.

CAMEO 21

Napier's inequality and the limit for e

Approximations to the area of a region bounded by an arc of the hyperbola $y = 1/x$ lead to inequalities for natural logarithms. For example, if we bound the region under the graph of the hyperbola over the interval $[a, b]$ with inscribed and circumscribed rectangles and compute areas, as shown in Figure 21.1, we obtain *Napier's inequality*, named for the Scottish mathematician John Napier (1550–1617), the inventor of logarithms:

$$\boxed{\text{if } 0 < a < b, \text{ then } \frac{1}{b} < \frac{\ln b - \ln a}{b - a} < \frac{1}{a}.} \qquad (21.1)$$

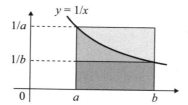

Figure 21.1. Illustrating Napier's inequality

Comparing areas of the rectangles and the region under the hyperbola yields

$$\frac{1}{b}(b-a) < \int_a^b \frac{1}{x} dx < \frac{1}{a}(b-a)$$

from which (21.1) follows.

In Cameo 12 we saw a derivation of $\lim_{n\to\infty}(1 + \frac{1}{n})^n = e$ using secant and tangent lines to curves of the form $y = b^x$. In this Cameo we can derive the limit using Napier's inequality. Setting $a = n$ and $b = n + 1$ in (21.1) yields

$$\frac{1}{n+1} < \ln\left(1 + \frac{1}{n}\right) < \frac{1}{n} \qquad (21.2)$$

and multiplication by n and exponentiation yields

$$e^{n/(n+1)} < \left(1 + \frac{1}{n}\right)^n < e, \qquad (21.3)$$

from which $\lim_{n\to\infty}(1 + \frac{1}{n})^n = e$ follows using the squeeze theorem.

Exercise 21.1. Show that $\lim_{n\to\infty}(1+\frac{1}{n})^{n+1} = e$. (Hint: first establish

$$e < \left(1+\frac{1}{n}\right)^{n+1} < e^{(n+1)/n} \qquad (21.4)$$

by multiplying (21.2) by $n+1$.)

Exercise 21.2. Show that for every $n \geq 1$, $(1+\frac{1}{n})^n < e < (1+\frac{1}{n})^{n+1}$.

Exercise 21.3. Show that $\lim_{x\to 1}(\ln x)/(x-1) = 1$. (Hint: set $(a,b) = (1,x)$ for $x > 1$ and $(a,b) = (x,1)$ for $0 < x < 1$ in (21.1).)

Exercise 21.4. Show that for $x > -1$ and $x \neq 0$, $x/(1+x) < \ln(1+x) < x$. See Figure 21.2. (Hint: consider two cases as in Exercise 21.3.)

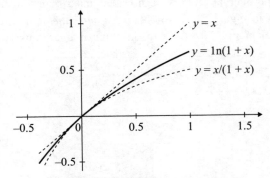

Figure 21.2. The inequality in Exercise 21.4

Taking reciprocals in (21.1) yields

$$\text{if } 0 < a < b, \text{ then } a < \frac{b-a}{\ln b - \ln a} < b,$$

where $(b-a)/(\ln b - \ln a)$ is the logarithmic mean of a and b from Cameo 10. In Cameo 29 we show that the logarithmic mean lies between the geometric and arithmetic means of a and b, also encountered in Cameo 10.

Exercise 21.5. Give a second proof of Napier's inequality by comparing slopes of the lines tangent to the graph of $y = \ln x$ at $(a, \ln a)$ and $(b, \ln b)$ and the secant line joining $(a, \ln a)$ and $(b, \ln b)$, as shown in Figure 21.3.

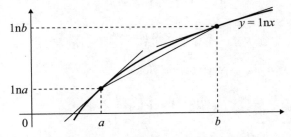

Figure 21.3. A second proof of Napier's inequality

CAMEO 21. Napier's inequality and the limit for *e*

John Napier

SOURCES: N. Schaumberger, "An alternate classroom proof of the familiar limit for *e*, *Two-Year College Mathematics Journal*, **3** (1972), pp. 72–73, and RBN, "Napier's inequality (two proofs)," *College Mathematics Journal*, **24** (1993), p. 165.

CAMEO 22

The nth root of $n!$ and another limit for e

The factorial function $n! = 1 \times 2 \times 3 \times \cdots \times n$ increases rapidly with n. The derivative of a function provides us with a way to measure how rapidly a function increases, but it only applies to certain functions defined on intervals. Since the domain of the factorial function is the set of nonnegative integers, we seek another approach. Comparing the nth root of $n!$ to n yields the data in Table 22.1.

Table 22.1. The nth root of n factorial

n	10	100	1000	10000	100000
$\sqrt[n]{n!}$	4.5287	37.9927	369.4916	3680.8272	36790.3999
$n/\sqrt[n]{n!}$	2.2081	2.6321	2.7064	2.7168	2.7181

In this Cameo we use integration to show that

$$\lim_{n \to \infty} \frac{n}{\sqrt[n]{n!}} = e. \qquad (22.1)$$

Exercise 22.1. Show that $\ln(\sqrt[n]{n!}/n) = \sum_{k=1}^{n} \ln(k/n) \cdot (1/n)$. (Hint: $\sqrt[n]{n!}/n = \sqrt[n]{n!/n^n}$.)

Exercise 22.2. The sum in the preceding exercise looks as if it might be a Riemann sum. Is it? If so, what definite integral does it approximate? (Hint: see Figure 22.1.)

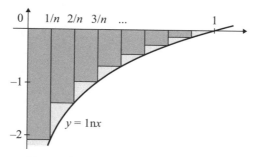

Figure 22.1. Illustrating a Riemann sum

Since the Riemann sum in Exercise 22.1 approximates the convergent improper integral $\int_0^1 \ln x \, dx$, we now find its value.

Exercise 22.3. Show that $\int_0^1 \ln x \, dx = -1$.

Exercise 22.4. Conclude that $\lim_{n \to \infty} (\sqrt[n]{n!}/n) = 1/e$, which is equivalent to (22.1).

SOURCE: Adapted from C. C. Mumma II, "$N!$ and the root test," *American Mathematical Monthly*, **93** (1986), p. 561.

CAMEO 23

Does shell volume equal disk volume?

Two applications of the definite integral are to find the volume of a solid of revolution by the so-called disk method and the cylindrical shell method. Both methods can be applied to some solids of revolution, which yields a convenient way for students to check their work: compute the volume both ways, the answers should agree. But is it true that both methods must agree?

We begin by considering the relationship between the definite integrals of a continuous monotone function $y = h(x)$ and its inverse $x = h^{-1}(y)$ on the interval $x \in [a, b]$, as illustrated in Figure 23.1a for an increasing function and in Figure 23.1b for a decreasing function (in the case where the common graph of the two functions lies in the first quadrant).

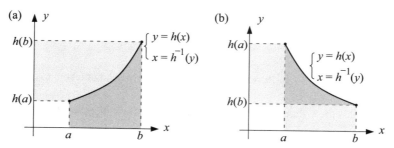

Figure 23.1. Graphs of $y = h(x)$ for h increasing and h decreasing

Area interpretations of integrals in Figure 23.1a yield

$$\int_a^b h(x)dx + \int_{h(a)}^{h(b)} h^{-1}(y)dy = bh(b) - ah(a)$$

or

$$\int_a^b h(x)dx = bh(b) - ah(a) - \int_{h(a)}^{h(b)} h^{-1}(y)dy. \tag{23.1}$$

Exercise 23.1. Show that (23.1) also holds in Figure 23.1b. (Hint: note that the integral $\int_{h(a)}^{h(b)} h^{-1}(y)dy$ represents the negative of the area of the region to the left of the graph of $x = h^{-1}(y)$ over the interval $y \in [h(b), h(a)]$.)

A rigorous proof of (23.1) requires Riemann sums, but if we make the additional assumption that $h'(x)$ is continuous, we can give a simple proof using integration by parts in $\int_a^b h(x)dx$.

With $u = h(x)$ and $dv = dx$ we obtain

$$\int_a^b h(x)dx = xh(x)|_a^b - \int_a^b xh'(x)dx = bh(b) - ah(a) - \int_a^b xh'(x)dx.$$

If we now substitute $y = h(x)$ so that $x = h^{-1}(y)$ and $h'(x)dx = dy$ in the rightmost integral, we obtain (23.1).

We now use (23.1) to prove the equivalence of the disk and shell methods for a region in the first quadrant bounded by the graph of an increasing function (other cases are similar).

Theorem. *Let f be continuous, increasing, and positive on $[a, b]$ where $0 \leq a < b$, and let R denote the region bounded by the graph of $y = f(x)$, the x-axis, and the lines $x = a$ and $x = b$. Let S denote the solid obtained by revolving R about the x-axis. Then the disk and shell methods for computing the volume of S yield the same result. See Figure 23.2.*

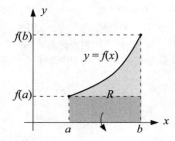

Figure 23.2. Revolving a region R to form a solid of revolution

If V_{shell} denotes the volume of S from the cylindrical shell method, then

$$V_{\text{shell}} = \pi[f(a)]^2(b-a) + \int_{f(a)}^{f(b)} 2\pi y[b - f^{-1}(y)]dy$$

$$= \pi b[f(a)]^2 - \pi a[f(a)]^2 + \pi by^2 \Big|_{f(a)}^{f(b)} - \int_{f(a)}^{f(b)} 2\pi y f^{-1}(y)dy$$

$$= \pi \left(b[f(b)]^2 - a[f(a)]^2 - \int_{f(a)}^{f(b)} 2y f^{-1}(y)dy \right). \tag{23.2}$$

Exercise 23.2. Show that we also obtain (23.2) by first computing the volume of a cylinder with height b and base radius $f(b)$ and then subtract both the volume of a cylinder with height a and base radius $f(a)$ and the volume (computed by the shell method) of the solid of revolution formed by rotating the region to the *left* of the graph of $y = f(x)$ about the x-axis.

The volume V_{disk} of S from the disk method is clearly $V_{\text{disk}} = \pi \int_a^b [f(x)]^2 dx$. Letting $h(x) = [f(x)]^2$ so that $h^{-1}(y) = f^{-1}(\sqrt{y})$ in (23.1) yields

$$V_{\text{disk}} = \pi \left(b[f(b)]^2 - a[f(a)]^2 - \int_{[f(a)]^2}^{[f(b)]^2} f^{-1}(\sqrt{y})dy \right).$$

CAMEO 23. Does shell volume equal disk volume?

If we substitute $t = \sqrt{y}$ in the last integral, $dy = 2t\,dt$ and thus

$$V_{\text{disk}} = \pi \left(b[f(b)]^2 - a[f(a)]^2 - \int_{f(a)}^{f(b)} 2t f^{-1}(t)\,dt \right) = V_{\text{shell}}$$

as desired.

SOURCE: E. Key, "Disks, shells, and integrals of inverse functions," *College Mathematics Journal,* **25** (1994), pp. 136–138.

CAMEO 24

Solids of revolution and the Cauchy-Schwarz inequality

Let f be a continuous nonnegative function on the interval $[a,b]$ where $0 < a < b$, and let R denote the region bounded by the graph of $y = f(x)$, the x-axis, $x = a$, and $x = b$. Now generate two solids, one by revolving R about the x-axis, the other by revolving R about the y-axis. See Figure 24.1.

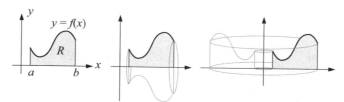

Figure 24.1. Revolving a region R about the x- and the y-axes

Using the disk method, the volume of the solid obtained when R is revolved about the x-axis is $V_{x\text{-axis}} = \int_a^b \pi[f(x)]^2 dx$, and using the shell method, the volume of the solid obtained when R is revolved about the y-axis is $V_{y\text{-axis}} = \int_a^b 2\pi x f(x) dx$. How are the volumes of the two solids related? In this Cameo we first show that

$$V_{y\text{-axis}}^2 \leq \frac{4}{3}\pi(b^3 - a^3) \cdot V_{x\text{-axis}}. \tag{24.1}$$

The fact that an inequality exists between the two volumes is somewhat surprising. Furthermore, the inequality itself is a bit strange, as the first term $\frac{4}{3}\pi(b^3 - a^3)$ in the product on the right doesn't depend on the function f, and it can be interpreted as the difference of the volumes of two spheres, one with radius b and another with radius a.

Exercise 24.1. Under the assumptions in the first paragraph, show that for any t,

$$\int_a^b \pi[f(x) + tx]^2 dx \geq 0.$$

Exercise 24.2. Show that the integral in Exercise 24.1 can be written as

$$\int_a^b \pi[f(x) + tx]^2 dx = At^2 + Bt + C, \tag{24.2}$$

where $A = \frac{1}{3}\pi(b^3 - a^3)$, $B = V_{y\text{-axis}}$, and $C = V_{x\text{-axis}}$.

71

Exercise 24.3. How are the coefficients A, B, and C related when $At^2 + Bt + C \geq 0$? (Hint: show that $At^2 + Bt + C \geq 0$ with $A > 0$ implies that $B^2 - 4AC \leq 0$.)

Exercise 24.4. Show that $B^2 - 4AC \leq 0$ is equivalent to (24.1).

Exercise 24.5. Show that (24.1) is a best-possible inequality—that is, there are functions f that yield equality in (24.1). (Hint: consider $f(x) = mx$ for $m > 0$.)

Exercise 24.6. Show that $V_{y\text{-axis}} - V_{x\text{-axis}} \leq \frac{\pi}{3}(b^3 - a^3)$. (Hint: let $t = -1$ in (24.2).) Is the inequality best-possible?

The same procedure outlined above in Exercises 24.2 through 24.4 can be used to prove the *Cauchy-Schwarz inequality* for definite integrals: If f and g are continuous on the interval $[a, b]$, then

$$\left[\int_a^b f(x)g(x)dx \right]^2 \leq \int_a^b [f(x)]^2 dx \cdot \int_a^b [g(x)]^2 dx. \tag{24.3}$$

Exercise 24.7. Prove (24.3). (Hint: consider $\int_a^b [f(x) + tg(x)]^2 dx \geq 0$.)

Exercise 24.8. Use the Cauchy-Schwarz inequality to show that if the region R in Figure 24.1 has area A_R and the average value of f on $[a, b]$ is f_{ave}, then

$$V_{x\text{-axis}} \geq \pi A_R f_{\text{ave}}.$$

(Hint: in (24.3) let $g(x) = 1$ and recall that $f_{\text{ave}} = A_R/(b-a)$.) Is this inequality best-possible? (Hint: let $f(x) = k > 0$.)

Exercise 24.9. Let $(\bar{x}, \bar{y})$ denote the centroid of the region R in Figure 24.1. Show that $\bar{y} \geq f_{\text{ave}}/2$. (Hint: $V_{x\text{-axis}} = 2\pi \bar{y} A_R$.)

The Cauchy-Schwarz inequality

The inequality (24.3)—also known as the *Cauchy-Bunyakovsky-Schwarz inequality*—is one of the most important in mathematics, and finds applications in a variety of areas including linear algebra, probability, and analysis. In 1821 Augustin-Louis Cauchy (1789–1857) published the version for sums. The integral version (24.3) was published in 1859 by Viktor Yakovlevich Bunyakovsky (1804–1889), and rediscovered in 1885 by Hermann Amandus Schwarz (1843–1921), whose proof was essentially the same as Exercise 24.7.

CAMEO 24. Solids of revolution and the Cauchy-Schwarz inequality

Cauchy, Bunyakovsky, and Schwarz

CAMEO 25

The midpoint rule is better than the trapezoidal rule

After deriving the formulas for the midpoint and trapezoidal rules, and before presenting the error bounds, we can use the following illustration to show that the midpoint rule is more accurate than the trapezoidal rule for continuous concave down functions. A similar illustration, which makes a good exercise, is to show that the same is true for continuous concave up functions. The question "how much better?" leads naturally to discussion of the bounds on the errors.

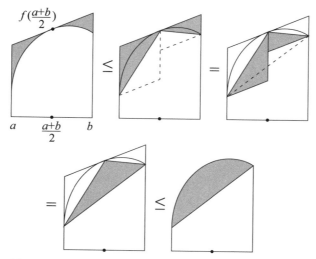

Figure 25.1. Comparing the midpoint and trapezoidal rules

Exercise 25.1. Show that the same result holds for continuous positive concave up functions.

SOURCE: F. Burk, "Behold! The midpoint rule is better than the trapezoidal rule for concave functions," *College Mathematics Journal*, **16** (1985), p. 56.

CAMEO 26

Can the midpoint rule be improved?

Let f be a continuous positive concave down function on an interval $[a, b]$, and consider the midpoint rule approximation M_1 to $\int_a^b f(x)\,dx$ using a single interval and midpoint. Most calculus texts illustrate the fact that M_1 is equal to the area of a trapezoid, i.e., the area under the line tangent to f at the midpoint $x = (a+b)/2$. See Figure 26.1a.

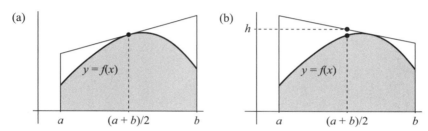

Figure 26.1. A modification of the midpoint rule

But is this particular trapezoid the best one to use? Does one constructed from another tangent line, such as the one illustrated in Figure 26.1b, perform better? The answer is easily seen to be no. As the shaded area under the graph of $y = f(x)$ is constant, the unshaded area in Figure 26.1b (the error in approximating $\int_a^b f(x)\,dx$ by the area under a tangent line) will be minimized when the total area below the tangent line is minimized. This area is the base $b-a$ times the height h of the tangent line at the midpoint $(a+b)/2$, so it suffices to minimize h. And that occurs when the point of tangency is at $x = (a+b)/2$.

Exercise 26.1. Show that the same result holds for continuous positive concave up functions.

Exercise 26.2. Show that M_1, usually defined as the area of a rectangle, is also the area of a trapezoid for continuous positive functions.

SOURCE: R. Paré, "A visual proof of Eddy and Fritsch's minimal area property," *College Mathematics Journal*, **26** (1995), pp. 43–44.

CAMEO 27

Why is Simpson's rule exact for cubics?

One of the great freebies of calculus is the fact that Simpson's rule, guaranteed to be exact for quadratics, is also exact for cubics. Of course this follows from the error bound, but the following explanation is more direct. It suffices to consider the interval $[-h, h]$ for arbitrary $h > 0$. Let f be an arbitrary cubic polynomial, and let g be the unique quadratic polynomial that agrees with f at $-h$, 0, and h. Now let $p(x) = f(x) - g(x)$. See Figure 27.1.

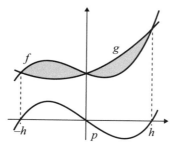

Figure 27.1. Graphs of typical functions f, g, and p

Since $p(-h) = p(0) = p(h) = 0$, the three zeros of p are $-h$, 0, and h. Therefore $p(x) = ax(x+h)(x-h) = ax^3 - ah^2 x$ (where a is the coefficient of x^3 in f). Hence p is an odd function, so that $\int_{-h}^{h} p(x)\,dx = 0$. Thus $\int_{-h}^{h} f(x)\,dx = \int_{-h}^{h} g(x)\,dx$, that is, the two gray regions in Figure 27.1 have the same area. Hence, since Simpson's rule is exact for g, it is also exact for f.

SOURCE: Adapted from R. N. Greenwell, "Why Simpson's rule gives exact answers for cubics," *Mathematical Gazette*, **83** (1999), p. 508.

CAMEO 29

The Hermite-Hadamard inequality

A surprisingly useful—and easy to prove—double inequality is the *Hermite-Hadamard inequality*, named for the French mathematicians Charles Hermite (1822–1901) who first published it in 1883, and Jacques Hadamard (1865–1963) who rediscovered it ten years later.

Charles Hermite and Jacques Hadmard.

The Hermite-Hadamard inequality. *If f is continuous and concave up on $[a,b]$, then*

$$f\left(\frac{a+b}{2}\right) \leq \frac{1}{b-a}\int_a^b f(x)\,dx \leq \frac{f(a)+f(b)}{2}, \qquad (29.1)$$

and if f is continuous and concave down on $[a,b]$, then

$$\frac{f(a)+f(b)}{2} \leq \frac{1}{b-a}\int_a^b f(x)\,dx \leq f\left(\frac{a+b}{2}\right). \qquad (29.2)$$

The middle term in the Hermite-Hadamard inequality is the *average value* f_{ave} of f on $[a,b]$, and the inequality states that f_{ave} lies between f evaluated at the average of a and b and the average of f at a and f at b.

To derive the inequality, consider the integral $\int_a^b f(x)\,dx$ and its trapezoidal rule approximation T_1 with one trapezoid and midpoint rule approximation M_1 with just one midpoint. See Figure 29.1 for the concave up case, when the value of the integral is larger than M_1 and smaller than T_1, so that

$$(b-a)f\left(\frac{a+b}{2}\right) \le \int_a^b f(x)\,dx \le (b-a)\frac{f(a)+f(b)}{2},$$

from which (29.1) follows. The proof for the concave down case is similar.

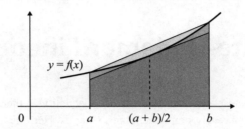

Figure 29.1. Proving the Hermite-Hadamard inequality

Example 29.1. As an application, consider the function $f(x) = \cos x$ on the interval $[-\pi/2, \pi/2]$. Since this function is concave down, (29.2) yields for a and b in $[-\pi/2, \pi/2]$

$$\frac{\cos a + \cos b}{2} \le \frac{\sin b - \sin a}{b-a} \le \cos\frac{a+b}{2}.$$

When $\{a,b\} = \{0,t\}$ for $t \ne 0$ in $[-\pi/2, \pi/2]$ we have

$$\frac{1+\cos t}{2} \le \frac{\sin t}{t} \le \cos\frac{t}{2}.$$

This double inequality is an improvement over the double inequality $\cos t \le (\sin t)/t \le 1$ discussed in Cameo 1, as can be seen in Figure 29.2.

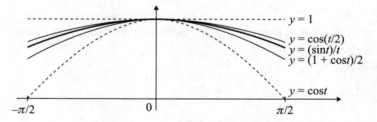

Figure 29.2. An illustration of the Hermite-Hadamard inequality

Exercise 29.1. Use (29.1) with $f(t) = 1/t$ and positive numbers 1 and $x \ne 1$ to derive a double inequality for $(\ln x)/(x-1)$. How does the result compare to the inequalities used in Exercise 21.1 to show that $\lim_{x\to 1}(\ln x)/(x-1) = 1$?

Example 29.2. Now consider the function $f(x) = e^x$ on the interval $[\ln a, \ln b]$ where $0 < a < b$. Since the graph of the function is concave up, (29.1) yields

$$e^{(\ln a + \ln b)/2} < \frac{e^{\ln b} - e^{\ln a}}{\ln b - \ln a} < \frac{e^{\ln a} + e^{\ln b}}{2}$$

CAMEO 28

Approximating π with integration

In Cameo 2 we approximated π using the limit of $n\sin(\pi/n)$ as $n \to \infty$ through powers of 2, and noted that geometrically we were finding the area of polygons inscribed in a circle of radius 1. We can do something similar with numeric integration. Since the area of one-quarter of a circle with radius 2 is π (see Figure 28.1a), we can approximate π with approximations to the definite integral $\int_0^2 \sqrt{4-x^2}\,dx$.

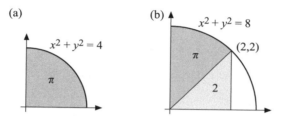

Figure 28.1. π as the area of two circular sectors

Example 28.1. Let's use Simpson's rule to approximate $\int_0^2 \sqrt{4-x^2}\,dx$. The results we obtain depend naturally on the software and the number n of subintervals in Simpson's rule. Here are some typical results where S_n denotes the Simpson's rule approximation with n subintervals:

Table 28.1.

n	S_n
10	3.1364470643
20	3.1397753524
40	3.1409504859
100	3.1414302492

The disappointing results (e.g., only three correct decimals with $n = 100$) are not surprising, since the first and higher order derivatives of the integrand are unbounded at $x = 2$ (recall the formula for the error bound for Simpson's rule).

Example 28.2. One-eighth of a circle of radius $\sqrt{8}$ also has area π, so perhaps using one-eighth of the circle $x^2 + y^2 = 8$ (see Figure 28.1b) will produce better results. The integral

$\int_0^2 \sqrt{8-x^2}\,dx$ equals $\pi + 2$, but π and $\pi + 2$ look exactly the same to the right of the decimal point. Using Simpson's rule to evaluate $\int_0^2 \sqrt{8-x^2}\,dx$ and subtracting 2 yields the following approximations to π.

Table 28.2.

n	$S_n - 2$
10	3.1415918322
20	3.1415926017
40	3.1415926503
100	3.1415926535

The derivatives of this integrand are bounded on [0,2], and consequently the results are much better. We get five decimals correct with $n = 10$; and ten decimals correct with $n = 100$.

Exercise 28.1. Show that

$$\pi = \int_0^1 \frac{6}{\sqrt{4-x^2}}\,dx = \int_0^1 \frac{4}{1+x^2}\,dx,$$

and use Simpson's rule with these integrals to approximate π. See Figure 28.2.

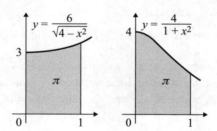

Figure 28.2. Two regions with area π

An ancient but common rational approximation to π is 22/7. In the next exercise we use calculus (rather than a calculator) to show that $\pi < 22/7$.

Exercise 28.2. Prove that $\pi < 22/7$. (Hint: show that

$$\frac{22}{7} - \pi = \int_0^1 \frac{x^4(1-x)^4}{1+x^2}\,dx.$$

This is problem A1 from the 1968 Putnam Competition discussed in Cameo 20.)

CAMEO 29. The Hermite-Hadamard inequality

(the inequalities are strict in this case) which is equivalent to the *arithmetic mean-logarithmic mean-geometric mean inequality*: for positive numbers a and b with $a \neq b$

$$\boxed{\sqrt{ab} < \frac{b-a}{\ln b - \ln a} < \frac{a+b}{2}.} \tag{29.3}$$

The inequality between the arithmetic and geometric means was established in Cameo 11, and the logarithmic mean was introduced in Exercise 10.1 and seen again in Cameo 21.

Exercise 29.2. Show that for n a positive integer,

$$\frac{1}{\sqrt{n(n+1)}} > \ln\left(1 + \frac{1}{n}\right) > \frac{1}{n + (1/2)}. \tag{29.4}$$

(Hint: set $a = n$ and $b = n+1$ in (29.3) and take reciprocals.)

The left-hand inequality in (29.4) is equivalent to $(1 + (1/n))^{\sqrt{n(n+1)}} < e$ and the right-hand inequality is equivalent to $e < (1 + (1/n))^{n+(1/2)}$, so that for every positive integer n,

$$\boxed{\left(1 + \frac{1}{n}\right)^{\sqrt{n(n+1)}} < e < \left(1 + \frac{1}{n}\right)^{n+(1/2)}.} \tag{29.5}$$

In Cameo 12 we established the inequality

$$\left(1 + \frac{1}{n}\right)^n < e < \left(1 + \frac{1}{n}\right)^{n+1}. \tag{29.6}$$

So how does the new boxed inequality (29.5) compare to the traditional one (29.6)? Here are some data with $n = 10$ and $n = 50$:

Table 29.1. A comparison of (29.5) and (29.6)

	Bounds on e	$n = 10$	$n = 50$
(29.6):	$(1 + 1/n)^n$	2.59374 (−4.48%)	2.69159 (−0.982%)
	$(1 + 1/n)^{n+1}$	2.85312 (+4.96%)	2.74542 (+0.998%)
(29.5):	$(1 + 1/n)^{\sqrt{n(n+1)}}$	2.71725 (−0.038%)	2.71824 (−0.0016%)
	$(1 + 1/n)^{n+1/2}$	2.72034 (+0.076%)	2.71837 (+0.0033%)

Note the remarkable improvement in percent relative error (in parentheses) for (29.5) over (29.6)!

CAMEO 30

Polar area and Cartesian area

Suppose we have a region in the plane for which it is possible to find its area using either Cartesian (i.e., xy-) coordinates or polar (i.e., $r\theta$-) coordinates. We certainly hope that the two procedures yield the same answer for the area, and in this Cameo we show, in a set of exercises, that they indeed do.

We consider only a simple case—when the region lies in the first quadrant and its curved boundary is the graph of a function in both Cartesian and polar coordinates, as illustrated in Figure 30.1. Other cases can be dealt with similarly.

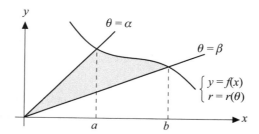

Figure 30.1. A region in Cartesian and polar coordinates

Let A_{cart} and A_{polar} denote the area of the shaded region in Figure 30.1 when computed in Cartesian and polar coordinates, respectively.

Exercise 30.1. Show that $A_{\text{cart}} = \int_a^b f(x)dx + \frac{1}{2}af(a) - \frac{1}{2}bf(b)$. (Hint: express the area of the shaded region in terms of the area under the graph of $y = f(x)$ and the areas of two right triangles.)

Exercise 30.2. Show that $A_{\text{cart}} = \frac{1}{2}[xf(x)|_b^a - 2\int_b^a f(x)dx]$. (Hint: notice the change in the order of the limits of integration in the integral.)

Exercise 30.3. Show that $A_{\text{polar}} = \frac{1}{2}\int_\beta^\alpha [r(\theta)]^2 d\theta = \frac{1}{2}\int_\beta^\alpha [r(\theta)\cos\theta]^2 d(\tan\theta)$.

Exercise 30.4. Show that $A_{\text{polar}} = \frac{1}{2}\int_b^a x^2 d(\frac{f(x)}{x})$. (Hint: use the change of variables $x = r(\theta)\cos\theta$ and $f(x) = y = r(\theta)\sin\theta$.)

Exercise 30.5. Show that $A_{\text{polar}} = \frac{1}{2}[x^2 \frac{f(x)}{x}|_b^a - \int_b^a \frac{f(x)}{x} \cdot 2x\,dx]$ to conclude that $A_{\text{polar}} = A_{\text{cart}}$. (Hint: integrate the result in Exercise 30.4 by parts and then compare to Exercise 30.2.)

87

For the final two exercises, consider the case where the function f has an inverse g, as illustrated in Figure 30.2, and set $A = f(a)$ and $B = f(b)$:

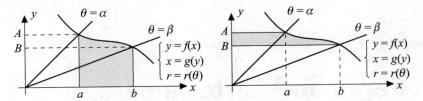

Figure 30.2. Two more regions in Cartesian and polar coordinates

Exercise 30.6. Derive the somewhat unexpected result that A_{polar} is equal to the average of the areas of the two shaded regions in Figure 30.2, i.e.,

$$A_{\text{polar}} = \frac{1}{2}\int_\beta^\alpha [r(\theta)]^2 d\theta = \frac{1}{2}\left(\int_a^b f(x)dx + \int_B^A g(y)dy\right).$$

(Hint: use the results in Exercises 30.1 and 30.5 to show that $A_{\text{polar}} = \int_B^A g(y)dy + \frac{1}{2}bf(b) - \frac{1}{2}af(a)$.)

Exercise 30.7. Are there any curves in the first quadrant for which the three shaded regions in Figures 30.1 and 30.2 have identical areas for every choice of (positive) a and b? (Hint: the answer is yes. Equating any two of the three yields $af(a) = bf(b)$, so that $xf(x)$ must be a (positive) constant.)

SOURCE: Exercises 30.6 and 30.7 are adapted from G. Strang, "Polar area is the average of strip areas," *American Mathematical Monthly*, **100** (1993), pp. 250–254.

CAMEO 31

Polar area as a source of antiderivatives

In this Cameo we will learn that interpreting a definite integral as the area of a region in polar coordinates may help us find an antiderivative of the integrand. The idea is based on the fundamental theorem: if you can use geometry to evaluate $F(x) = \int_a^x f(\theta)d\theta$, then you have found an antiderivative of f since $\frac{d}{dx}F(x) = f(x)$.

Example 31.1. Using either a double angle formula or integration by parts, it is easy to show that $\int \cos^2 x \, dx = (1/2)(x + \sin x \cos x) + C$. Now consider a definite integral in polar coordinates with the same integrand, i.e., $\int_0^\alpha \cos^2 \theta \, d\theta$. This integral looks suspiciously like a polar area integral since the integrand is a square. Since the graph of $r = 2\cos\theta$ is a circle, we write $\int_0^\alpha \cos^2 \theta \, d\theta = \frac{1}{2} \cdot \frac{1}{2} \int_0^\alpha (2\cos\theta)^2 \, d\theta$ so that $\int_0^\alpha \cos^2 \theta \, d\theta$ represents one-half the area of the region in the plane bounded by $r = 2\cos\theta$ and the rays $\theta = 0$ and $\theta = \alpha$, as illustrated in Figure 31.1 for α in $(0, \pi/2)$. (The coordinates for P in the figure are its polar coordinates, the Cartesian coordinates of P are $(2\cos^2\alpha, 2\cos\alpha\sin\alpha)$.)

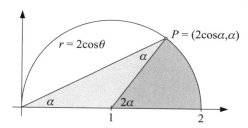

Figure 31.1. An area interpretation of $\int_0^\alpha \cos^2 \theta \, d\theta$

We can also compute the area of the region using geometry. The circular sector (in dark gray) has angle 2α and radius 1, so its area is $(1/2)(2\alpha) = \alpha$. The light gray triangle has base 1 and altitude $2\cos\alpha\sin\alpha$ (the y-coordinate of P) so its area is $\sin\alpha\cos\alpha$. Hence

$$\int_0^\alpha \cos^2 \theta \, d\theta = \frac{1}{2}(\alpha + \sin\alpha\cos\alpha) = \frac{1}{2}(\theta + \sin\theta\cos\theta)\big|_0^\alpha,$$

and thus

$$\int \cos^2 \theta \, d\theta = \frac{1}{2}(\theta + \sin\theta\cos\theta) + C.$$

Exercise 31.1. Use this polar area method to integrate the square of the secant by showing that $\int_0^\alpha \sec^2\theta\,d\theta = \tan\alpha$. (Hint: consider the graph of $r = \sec\theta$.)

Exercise 31.2. Show that

$$\int \frac{d\theta}{(a\cos\theta + b\sin\theta)^2} = \frac{1}{a(a\cot\theta + b)} + C.$$

(Hint: the graph of $r = 1/(a\cos\theta + b\sin\theta)$ is a straight line! Sketch a picture for a and b both positive.)

Exercise 31.3. Show that

$$\int \frac{d\theta}{(1+\cos\theta)^2} = \frac{\sin\theta(2+\cos\theta)}{3(1+\cos\theta)^2} + C.$$

(Hint: the graph of $r = 1/(1+\cos\theta)$ is a parabola, so you will need to use geometry and integration in Cartesian coordinates to find the area of the region.)

CAMEO 32

The prismoidal formula

A *prismatoid* is a polyhedron all of whose vertices lie in two parallel planes, as illustrated in Figure 32.1.

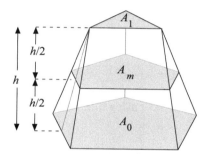

Figure 32.1. A prismatoid

The volume V of a prismatoid can be found using the *prismoidal formula*. Let A_0 and A_1 be the areas of the faces in the two parallel planes, and let A_m be the area of the intersection of the prismatoid with the plane parallel to the two planes and midway between them. If h denotes the distance between the planes with areas A_0 and A_1, then

$$V = \frac{h}{6}(A_0 + 4A_m + A_1). \tag{32.1}$$

Example 32.1. A pyramid with a square base is a prismatoid. If the length of a side of the base is s and the height h, then $A_0 = s^2$, $A_1 = 0$, and $A_m = (s/2)^2$, so that $V = (h/6)(0 + s^2 + s^2) = (1/3)s^2 h$, the familiar formula for the volume of a pyramid.

In this Cameo we examine whether the prismoidal formula applies to solids other than prismatoids.

Example 32.2. Does the prismoidal formula work for a sphere? Let r be the radius of the sphere, and let the two planes be tangent to the sphere at the north and south poles. Then $A_0 = A_1 = 0$, $A_m = \pi r^2$ is the area of the circle at the equator of the sphere, and $h = 2r$. Then (32.1) yields $V = (2r/6)(0 + 4\pi r^2 + 0) = (4/3)\pi r^3$, the correct volume of a sphere.

Exercise 32.1. Is the prismoidal formula exact for cylinders and cones?

You may have been surprised to learn that the prismoidal formula gives the exact volume for cylinders and cones, as well as for spheres. It also gives the exact volumes for many other solids. Let's see why.

Example 32.3. Using the disk method, let's set up an integral of the form $\int_a^b A(x)dx$ for the volume of a sphere (where $A(x)$ is the area of the cross-section at the point x between a and b). To obtain a sphere of radius r using the disk method we revolve the gray semicircle bounded by the graph of $y = \sqrt{r^2 - x^2}$ and the x-axis for x in $[-r, r]$ about the x-axis, as shown in Figure 32.2:

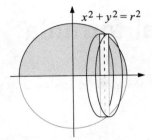

Figure 32.2. The volume of a sphere using the disk method

The cross-sectional area is $A(x) = \pi(\sqrt{r^2-x^2})^2 = \pi(r^2-x^2)$, and thus the volume of the sphere is $V_{\text{sphere}} = \int_{-r}^{r} \pi(r^2 - x^2)dx$. Rather than evaluate the integral using the fundamental theorem of calculus, let's approximate it using Simpson's rule with two subintervals, S_2. Since $\Delta x = 2r/2 = r$, we have

$$V_{\text{sphere}} \approx S_2 = \frac{r}{3}[A(-r) + 4A(0) + A(r)] = \frac{r}{3}[0 + 4\pi r^2 + 0] = \frac{4}{3}\pi r^3.$$

Observe two things: S_2 *is* the prismoidal formula, and it is *exact*. In fact, the prismoidal formula will be exact whenever S_2 is exact, as it is for solids whose cross-sectional area function is a polynomial of degree 3 or less (as shown in Cameo 27).

The *Moscow Papyrus*, an ancient Egyptian papyrus dating from about 1850 BCE, contains 25 mathematical problems. The 14th problem concerns the volume of a frustum of a pyramid, as seen in Figure 31.3a. A *frustum* (Latin for "a piece") of a pyramid is a portion of the pyramid lying between two planes parallel to the base, as illustrated in Figure 32.3b.

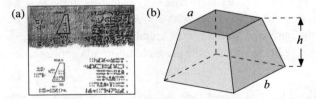

Figure 32.3. A portion of the Moscow papyrus and a frustum of a pyramid

For a frustum with square bases measuring a and b on a side and height h, the Papyrus gives the volume V as (in modern notation)

$$V = \frac{h}{3}(a^2 + ab + b^2).$$

Exercise 32.2. Is this ancient formula a special case of the prismoidal formula? Is it exact?

PART III
Infinite Series

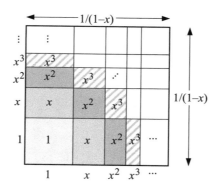

CAMEO 33

The geometry of geometric series

Geometric series—series of the form $a + ar + ar^2 + \cdots + ar^n + \cdots$ for nonzero a and r—are among the first encountered in calculus, and many of them can be illustrated visually.

Example 33.1. An illustration of the geometric series with $a = r = 1/2$ begins with a square with area 1. Cut it in half vertically, as shown on the left in Figure 33.1, to create two rectangles each with area $1/2$.

Figure 33.1. Dissecting a square with area 1

Then cut the right hand rectangle in half, to create two squares each with area $1/4$. Next cut the square in the upper right hand corner in half vertically, creating two rectangles each with area $1/8$. If we continue this process indefinitely we will have cut the square into infinitely many pieces whose total area is the same as the area of the original square, so that

$$\frac{1}{2} + \frac{1}{4} + \frac{1}{8} + \frac{1}{16} + \cdots = 1.$$

Example 33.2. Figure 33.2 shows a second dissection of a square with area 1. The white squares have areas $1/4$, $1/16$, $1/64$, and so on, as do the light gray and the dark gray squares. Thus

$$\frac{1}{4} + \frac{1}{16} + \frac{1}{64} + \cdots = \frac{1}{3}.$$

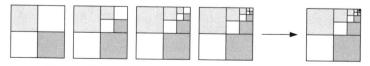

Figure 33.2. A second dissection of a square with area 1

95

Exercise 33.1. In Figure 33.3 we see two dissections of an equilateral triangle with area 1. What geometric series do they illustrate?

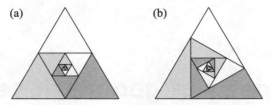

Figure 33.3. Two dissections of an equilateral triangle with area 1

When r is negative we can illustrate the geometric series by adding and subtracting areas of regions, as in the next example.

Example 33.3. In Figure 33.4 we show that the geometric series with $a = 1$ and $r = -1/2$ has sum $2/3$, beginning with an isosceles trapezoid with area 1 and alternately adding and subtracting the areas of smaller similar trapezoids.

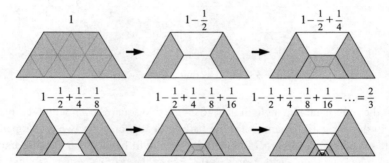

Figure 33.4. A dissection of a trapezoid

Exercise 33.2. What geometric series (with $r < 0$) is illustrated by the dissection of a square with area 1 in Figure 33.5?

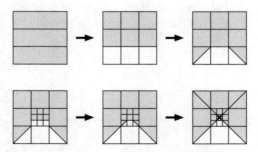

Figure 33.5. Another dissection of a square with area 1

Example 33.4. More generally, the sum $a/(1-r)$ of a geometric series with $a > 0$ and $0 < r < 1$ can be illustrated with rectangles as shown in Figure 33.6. The rectangle labeled

CAMEO 33. The geometry of geometric series

"a" has area a since its height is $a/(1-r)$ and its base is $1-r$, the rectangle labeled "ar" has area ar since its height is $a/(1-r)$ and its base is $r - r^2 = r(1-r)$, and so on. The entire gray rectangle has area $a/(1-r)$, thus illustrating $a + ar + ar^2 + \cdots = a/(1-r)$.

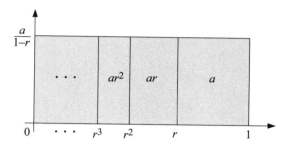

Figure 33.6. Summing a geometric series geometrically

Exercise 33.3. Use Figure 33.7 to show that $1 + r + r^2 + \cdots = 1/(1-r)$. (Hint: $\triangle PQR$ and $\triangle PST$ are similar triangles.)

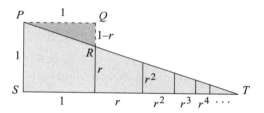

Figure 33.7. A right triangular representation of a geometric series

Convergence of an infinite series is defined in terms of the sequence $\{S_n\}$ of partial sums. For a geometric series with first term a and common ratio r, S_n has the form $S_n = a + ar + ar^2 + \cdots + ar^n$. In Figure 33.6, S_n is represented by the area of a rectangle with height $a/(1-r)$ and base $1 - r^{n+1}$ so that $S_n = [a/(1-r)] \cdot (1 - r^{n+1}) = a(1 - r^{n+1})/(1-r)$.

Since the geometry in Figures 33.6 and 33.7 only applies for positive values of r, we still need an algebraic argument to establish the closed formula for S_n. Perhaps the simplest way to do this is to compute S_{n+1} from S_n in two different ways. One way is to add the next term ar^{n+1} to S_n, i.e., $S_{n+1} = S_n + ar^{n+1}$, while a second way is to multiply S_n by r and add a: $S_{n+1} = a + rS_n$. These two ways to compute S_{n+1} yield the same number, hence

$$S_n + ar^{n+1} = a + rS_n, \text{ or } (1-r)S_n = a(1 - r^{n+1}),$$

and thus for $r \neq 1$ we have

$$S_n = a \frac{(1 - r^{n+1})}{(1-r)}$$

(when $r = 1$, S_n is simply $(n+1)a$).

See Example 17.4 in Cameo 17 for another method to find partial sums of geometric series.

Example 33.5. The recurrence $S_{n+1} = a + rS_n$ as well as the limit $S = a/(1-r)$ of the sequence $\{S_n\}$ is illustrated in Figure 33.8 for $a > 0$ and $0 < r < 1$. The coordinates of the black dots on the dashed line are (S_n, S_{n+1}).

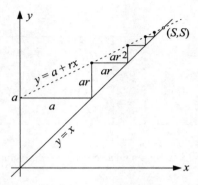

Figure 33.8. The convergence of $\{S_n\}$ to $S = a/(1-r)$ when $0 < r < 1$

When $-1 < r < 0$ the illustration looks like Figure 33.9.

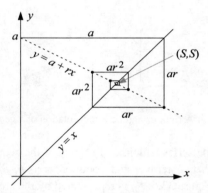

Figure 33.9. The convergence of $\{S_n\}$ to $S = a/(1-r)$ when $-1 < r < 0$

SOURCES:

Figure 33.5 is from H. Unal, "Proof without words: Sum of an infinite series," *College Mathematics Journal*, **40** (2009), p. 39.

Figure 33.6 is adapted from C. G. Spaht and C. M. Johnson, "Mathematics without words," *College Mathematics Journal*, **32** (2001), p. 109.

Figure 33.7 is from I. C. Bivens and B. G. Klein, "Geometric series," *Mathematics Magazine*, **61** (1988), p. 219.

Figures 33.8 and 33.9 are from The Viewpoints 2000 Group, "Proof without words: Geometric series," *Mathematics Magazine*, **74** (2001), p. 320.

CAMEO 34

Geometric differentiation of geometric series

As noted in the preceding Cameo, some of the first series studied in calculus are the geometric series: for real numbers a and r with $|r| < 1$ we have

$$a + ar + ar^2 + \cdots = \frac{a}{1-r}.$$

Consider the special case $a = 1$ and $r = x$ with x in $(0, 1)$, and construct a square with sides $1 + x + x^2 + \cdots = 1/(1-x)$ as shown in Figure 34.1, and use the terms of the series to partition the square into smaller squares and rectangles.

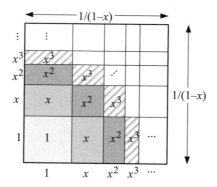

Figure 34.1. A geometric series partition of a square

Summing the areas of the interior squares and rectangles yields

$$1 + 2x + 3x^2 + 4x^3 + \cdots = \frac{1}{(1-x)^2},$$

which we may be tempted to write as the equivalent statement

$$\frac{d}{dx}(1 + x + x^2 + x^3 + \cdots) = \frac{d}{dx}\frac{1}{1-x}.$$

This *suggests* (but does not *prove*) that we may be able to differentiate each term in a convergent geometric series, and the series of derivatives will converge to the derivative of the sum of the original series. Series where the terms involve powers of a variable x are *power series*, and the study of the calculus of power series is a major focus of the series chapter in calculus texts.

Exercise 34.1. What series (and its sum) is represented by Figure 34.2? Does it represent the derivative of some other series?

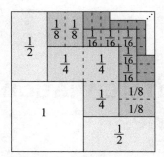

Figure 34.2. Another series partition of a square

SOURCE: Figure 34.1 is from RBN, "Mathematics without words," *College Mathematics Journal*, **32** (2001), p. 267.

CAMEO 35

Illustrating a telescoping series

Many calculus texts introduce telescoping series with the series

$$\frac{1}{1\cdot 2}+\frac{1}{2\cdot 3}+\frac{1}{3\cdot 4}+\cdots+\frac{1}{n(n+1)}+\cdots$$

and use partial fractions to show that the sequence of partial sums (and hence the series) converges to 1.

Example 35.1. We can illustrate the above result with the graphs of $y=x^{n-1}$ for x in $[0,1]$ and $n=1,2,3,\ldots$ shown in Figure 35.1 for $n=1,2,\ldots,9$. The areas of the regions between the graphs of $y=x^{n-1}$ and $y=x^n$ for $n=1,2,3,\cdots$ are the terms in the series, since

$$\int_0^1 (x^{n-1}-x^n)\,dx = \left[\frac{x^n}{n}-\frac{x^{n+1}}{n+1}\right]_0^1 = \frac{1}{n}-\frac{1}{n+1}=\frac{1}{n(n+1)}.$$

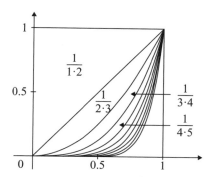

Figure 35.1. The telescoping series $\sum_{n=1}^{\infty} 1/n(n+1)$

It now follows that the nth partial sum S_n is

$$S_n = \frac{1}{1\cdot 2}+\frac{1}{2\cdot 3}+\frac{1}{3\cdot 4}+\cdots+\frac{1}{n(n+1)} = 1-\int_0^1 x^n\,dx = 1-\frac{1}{n+1}$$

and thus the sum of the series is $S=\lim_{n\to\infty} S_n = 1$.

See Example 17.5 in Cameo 17 for another method to evaluate the partial sums of this telescoping series.

Example 35.2. For another illustration of this series see Figure 35.2, where we use the graph of $y = 1/x$. The rectangle enclosing the portion of the graph over the interval $[n, n+1]$ has area $1 \cdot (\frac{1}{n} - \frac{1}{n+1}) = \frac{1}{n(n+1)}$, and the rectangles telescope to yield a rectangle with area 1.

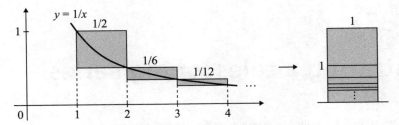

Figure 35.2. A second illustration of $\sum_{n=1}^{\infty} 1/n(n+1)$

Exercise 35.1. Show that
$$\frac{1}{1} + \frac{1}{1+2} + \frac{1}{1+2+3} + \cdots + \frac{1}{1+2+\cdots+n} + \cdots = 2.$$
(Hint: see Example 16.1).

SOURCE: Figure 35.1 is from J. H. Mathews, "The sum is one," *College Mathematics Journal*, **22** (1991), p. 322.

CAMEO 36

Illustrating applications of the monotone sequence theorem

In studying infinite series, an important tool for studying the convergence of sequences (in particular, sequences of partial sums) is the theorem known as

The monotone sequence theorem. *Bounded monotone sequences converge.* In particular, increasing sequences bounded above converge and decreasing sequences bounded below converge.

Example 36.1. As an illustration of the use of this theorem, consider the series

$$\sum_{k=1}^{\infty} \frac{1}{k^2} = 1 + \frac{1}{4} + \frac{1}{9} + \cdots + \frac{1}{n^2} + \cdots.$$

Many calculus texts use the integral test to show that this series converges. However, we will use the monotone sequence theorem to establish convergence. Since the terms of the series are positive, the sequence of partial sums is an increasing sequence. So we need only show that the partial sums are bounded above in order to invoke the monotone sequence theorem to conclude that the sequence of partial sums converges (and hence the series converges).

We represent each term of the series by the area of a square with side length $1/k$, and show that for any n, the sum of the first n terms of the series is less than 2. To do so, we place the squares inside a rectangle with height 1 and base 2 (recall that the sum of the geometric series with first term 1 and common ratio $1/2$ is 2). See Figure 36.1.

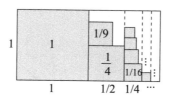

Figure 36.1. $1 + (1/4) + \cdots + (1/n^2)$ is bounded above by 2

In the column of squares above the interval on the base with length $1/2$ we can stack two squares, since the sum of their heights is less than 1. Similarly, in the column above the interval on the base with length $1/4$ we can stack four squares, since the sum of their heights is less than $4(1/4) = 1$. In general, in the column above the interval on the base of the rectangle with length $(1/2)^k$ we can stack 2^k squares, since the sum of their heights will be less than $2^k \cdot (1/2)^k = 1$. Hence for any n, the nth partial sum is less than 2. Using the Maclaurin series for the arcsine, it can be shown that the sum of this series is $\pi^2/6 \approx 1.645$.

Exercise 36.1. Modify Example 36.1 to show that the series

$$\sum_{k=1}^{\infty} \frac{1}{k^3} = 1 + \frac{1}{8} + \frac{1}{27} + \cdots + \frac{1}{n^3} + \cdots$$

converges. (Hint: see Figure 36.2.)

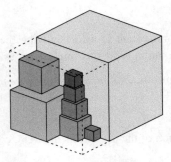

Figure 36.2. Bounding $1 + (1/8) + \cdots + (1/n^3)$

Example 36.2. For another illustration, we combine the monotone sequence theorem with the weighted AM-GM inequality from Example 11.5 in Cameo 11 to show that the sequence $\{[1 + (1/n)]^n\}_{n=1}^{\infty}$ often used to define e actually converges.

The weighted AM-GM inequality states that if a and b are positive numbers and $0 < r < 1$, then

$$a^r b^{1-r} \leq ra + (1-r)b \tag{36.1}$$

with equality if and only if $a = b$. See Figure 36.3 for an illustration.

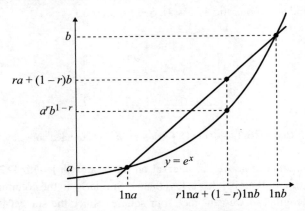

Figure 36.3. The weighted AM-GM inequality

CAMEO 36. Illustrating applications of the monotone sequence theorem

To show that $\{[1 + (1/n)]^n\}_{n=1}^{\infty}$ is increasing we set $a = 1$, $b = 1 + (1/n)$, and $r = 1/(n+1)$ in (36.1), yielding

$$\left(1 + \frac{1}{n}\right)^{n/(n+1)} < \frac{1}{n+1} \cdot 1 + \frac{n}{n+1} \cdot \frac{n+1}{n} = 1 + \frac{1}{n+1} \text{ so that}$$

$$\left(1 + \frac{1}{n}\right)^n < \left(1 + \frac{1}{n+1}\right)^{n+1}.$$

To show that $\{[1 + (1/n)]^n\}_{n=1}^{\infty}$ is bounded above, we set $a = 1$, $b = 1/2$, and $r = (n-1)/(n+1)$ in (36.1), yielding

$$\left(\frac{1}{2}\right)^{2/(n+1)} < \frac{n-1}{n+1} \cdot 1 + \frac{2}{n+1} \cdot \frac{1}{2} = \frac{n}{n+1} \text{ so that } \left(1 + \frac{1}{n}\right)^n < 4^{n/(n+1)} < 4.$$

Since $\{[1 + (1/n)]^n\}_{n=1}^{\infty}$ is increasing and bounded above, it converges by the monotone convergence theorem.

Exercise 36.2. Show that the sequence $\{[1 + (1/n)]^{n+1}\}_{n=1}^{\infty}$ converges by showing that the terms of the sequence are decreasing and bounded below. (Hint: use $a = 1$, $b = n/(n+1)$, $r = 1/(n+2)$ and note that the terms are all positive.)

Consequently, if $\lim_{n\to\infty}[1 + (1/n)]^n = e$, then we also have

$$\lim_{n\to\infty}\left(1 + \frac{1}{n}\right)^{n+1} = \lim_{n\to\infty}\left(1 + \frac{1}{n}\right) \cdot \lim_{n\to\infty}\left(1 + \frac{1}{n}\right)^n = 1 \cdot e = e,$$

and for any $n \geq 1$,

$$\left(1 + \frac{1}{n}\right)^n < e < \left(1 + \frac{1}{n}\right)^{n+1}.$$

In the next Cameo we present another example of the use of the monotone sequence theorem.

SOURCES:

Example 36.1 is adapted from M. K. Kinyon, "Another look at some *p*-series," *College Mathematics Journal*, **37** (2006), pp. 385–386, and G. Kimble, "Euler's other proof," *Mathematics Magazine*, **60** (1987), p. 282.

Figure 36.3 is from M. K. Brozinsky, "Proof without words," *College Mathematics Journal*, **25** (1994), p. 98.

Example 36.2 and Exercise 36.2 are adapted from N. S. Mendelsohn, "An application of a famous inequality," *American Mathematical Monthly*, **58** (1951), p. 563.

CAMEO 37

The harmonic series and the Euler-Mascheroni constant

One of the most important infinite series in calculus is the harmonic series. It is usually the first infinite series that the students encounter where the terms converge to zero yet the series diverges. Many calculus texts present Nicole Oresme's proof: if $H_n = 1 + \frac{1}{2} + \frac{1}{3} + \cdots + \frac{1}{n}$ denotes the nth partial sum, show that $H_{2^n} > 1 + \frac{n}{2}$ via a clever grouping of the terms. Some texts present the simple visual proof, where we interpret each term $1/k$ of the series as the area of a rectangle with base 1 and height $1/k$, and compare the nth partial sum H_n to the area under the graph of $y = 1/x$ over the interval $[1, n+1]$. See Figure 37.1 (the number above each shaded rectangle is its area). This proof has a pedagogical advantage of foreshadowing the integral test that usually follows the introduction of the harmonic series.

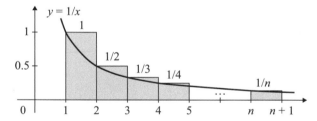

Figure 37.1. A partial sum of the harmonic series

Thus

$$H_n = \sum_{k=1}^{n} \frac{1}{k} > \int_1^{n+1} \frac{1}{x}\,dx = [\ln x]_1^{n+1} = \ln(n+1).$$

Since $\lim_{n\to\infty} \ln(n+1) = \infty$, we have $\lim_{n\to\infty} H_n = \infty$ and hence the harmonic series diverges. But it diverges very slowly. With the aid of a calculator or computer, it is easy to compute some values of H_n as shown in Table 37.1.

Table 37.1. Partial sums of the harmonic series

n	10	100	1,000	10,000	100,000	1,000,000
H_n	2.9290	5.1874	7.4855	9.7876	12.0901	14.3927

107

In Figure 37.1 we see that for each $n \geq 1$, the partial sum $H_n = 1 + \frac{1}{2} + \frac{1}{3} + \cdots + \frac{1}{n}$ is larger than $\ln(n+1)$. *But how much larger?* The answer to this question leads to a real number known as the *Euler-Mascheroni constant*. To compare H_n to $\ln(n+1)$, we add two rows to Table 37.1, as shown in Table 37.2:

Table 37.2. Comparing H_n to $\ln(n+1)$

n	10	100	1,000	10,000	100,000	1,000,000
H_n	2.9290	5.1874	7.4855	9.7876	12.0901	14.3927
$\ln(n+1)$	2.3979	4.6151	6.9088	9.2104	11.5129	13.8155
$H_n - \ln(n+1)$	0.5311	0.5723	0.5767	0.5772	0.5772	0.5772

If we set $\gamma_n = H_n - \ln(n+1)$ (with γ_0 equal to 0), then it appears that the sequence $\{\gamma_n\}$ in the last row is converging. We can now prove that $\{\gamma_n\}$ converges by showing that it is increasing and bounded above, using the monotone sequence theorem from Cameo 36. Observe that γ_n is the sum of the areas of the portions of the rectangles representing H_n that lie above the curve $y = 1/x$ over the interval $[1, n+1]$. See Figure 37.2 where γ_n appears in gray.

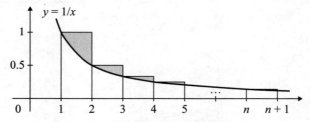

Figure 37.2. A visual representation of γ_n

To show that $\{\gamma_n\}$ is increasing, note that $\gamma_n - \gamma_{n-1}$ represents the area of the region shaded dark gray in Figure 37.3 and thus $\gamma_n - \gamma_{n-1} > 0$ so that $\gamma_n > \gamma_{n-1}$.

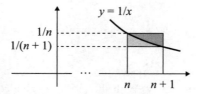

Figure 37.3. A visual representation of $\gamma_n - \gamma_{n-1}$

To show that $\{\gamma_n\}$ is bounded above, observe that $\gamma_n - \gamma_{n-1}$ is less than the area of the rectangle in two shades of gray in Figure 37.3, or

$$\gamma_n - \gamma_{n-1} < \frac{1}{n} - \frac{1}{n+1} = \frac{1}{n(n+1)}. \tag{37.1}$$

CAMEO 37. The harmonic series and the Euler-Mascheroni constant

Now replace n by k and sum both sides of (37.1) from 1 to n to obtain

$$\gamma_n = \gamma_n - \gamma_0 = \sum_{k=1}^{n}(\gamma_k - \gamma_{k-1}) < \sum_{k=1}^{n} 1/n(n+1).$$

Recall from Cameo 35 that $\sum_{k=1}^{n} 1/n(n+1)$ is a partial sum of a positive term series that telescopes to 1, hence $\gamma_n < 1$. Thus the sequence $\{\gamma_n\}$ converges, and its limit is traditionally denoted by the Greek letter γ:

$$\gamma = \lim_{n \to \infty} \gamma_n = \lim_{n \to \infty}[H_n - \ln(n+1)].$$

This limit is known as the *Euler-Mascheroni constant* after the Swiss mathematician Leonhard Euler (1707–1783) and the Italian mathematician Lorenzo Mascheroni (1750–1800). Evaluated to 20 decimal places, $\gamma \approx 0.57721566490153286060$. It is still unknown whether γ is rational or irrational.

We conclude this Cameo with another proof that the harmonic series diverges. This proof is attributed to the Italian mathematician Pietro Mengoli (1625–1686), and is based on the inequality

$$\frac{1}{n-1} + \frac{1}{n} + \frac{1}{n+1} > \frac{3}{n} \qquad (37.2)$$

for any $n \geq 2$, which follows from the concavity of the graph of $y = 1/x$. See Figure 37.4.

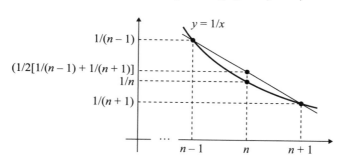

Figure 37.4. Proving Mengoli's inequality

In Figure 37.4 we see that

$$\frac{1}{2}\left(\frac{1}{n-1} + \frac{1}{n+1}\right) > \frac{1}{n} \quad \text{or} \quad \frac{1}{n-1} + \frac{1}{n+1} > \frac{2}{n},$$

which is equivalent to (37.2). Hence if H_n again denotes the nth partial sum of the harmonic series, then (37.2) implies that

$$H_{3n+1} = 1 + \left(\frac{1}{2} + \frac{1}{3} + \frac{1}{4}\right) + \left(\frac{1}{5} + \frac{1}{6} + \frac{1}{7}\right) + \cdots + \left(\frac{1}{3n-1} + \frac{1}{3n} + \frac{1}{3n+1}\right)$$

$$> 1 + \frac{3}{3} + \frac{3}{6} + \cdots + \frac{3}{3n} = 1 + H_n. \qquad (37.3)$$

Thus if the harmonic series converges to a real number H, then taking the limit as $n \to \infty$ of both sides of (37.3) yields the contradiction $H = 1 + H$, so the harmonic series must diverge.

Lorenzo Mascheroni, Leonhard Euler, and Pietro Mengoli.

CAMEO 38

The alternating harmonic series

Perhaps the simplest series to show convergent by the alternating series test (which we consider in the next Cameo) is the alternating harmonic series

$$1 - \frac{1}{2} + \frac{1}{3} - \cdots + (-1)^{n+1}\frac{1}{n} + \cdots = \sum_{n=1}^{\infty} (-1)^{n+1}\frac{1}{n}. \qquad (38.1)$$

In this Cameo we present a visual argument that it converges to $\ln 2$ by interpreting the terms in the series as areas of rectangles and $\ln 2$ as the area under the graph of $y = 1/x$ over the interval $[1, 2]$, as seen in Figure 38.1.

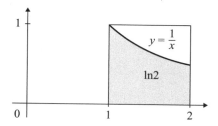

Figure 38.1. $\ln 2$ as the area of a portion of a square

In Figure 38.2a we see the first partial sum, 1, the area of a square with side 1 (the gray curve in the square is the graph of $y = 1/x$ over the interval $[1, 2]$). We now delete a rectangle with area $1/2$ (the right half of the square in Figure 38.2a) and add a rectangle with area $1/2 \cdot 2/3 = 1/3$, yielding the partial sum $1 - 1/2 + 1/3$ seen in Figure 38.2b.

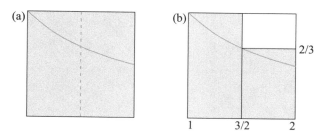

Figure 38.2. Two partial sums of (38.1)

Next we delete a rectangle with area $1/4$ and add a rectangle with area $1/4 \cdot 4/5 = 1/5$, then delete a rectangle with area $1/4 \cdot 2/3 = 1/6$ and add a rectangle with area $1/4 \cdot 4/7 = 1/7$, yielding the partial sum $1 - 1/2 + 1/3 - 1/4 + 1/5 - 1/6 + 1/7$ seen in Figure 38.3a.

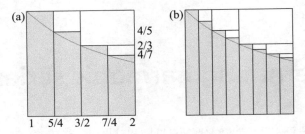

Figure 38.3. Two more partial sums of (38.1)

We continue in this manner with four more deletions and additions yielding the partial sum $1 - 1/2 + 1/3 - \cdots - 1/14 + 1/15$ in Figure 38.3b. In the limit we obtain the region (with area $\ln 2$) under the graph of $y = 1/x$ over the interval $[1, 2]$, as seen in Figure 38.1.

SOURCE: M. Hudelson, "Proof without words: The alternating harmonic series sums to $\ln 2$," *Mathematics Magazine*, **83** (2010), p. 294.

CAMEO 39

The alternating series test

Let $a_1 - a_2 + a_3 - \cdots + (-1)^{n+1} a_n + \cdots$ be a series of real numbers. If we let $S_{2k} = a_1 - a_2 + a_3 - a_4 + \cdots - a_{2k}$ and $S_{2k+1} = a_1 - a_2 + a_3 - \cdots - a_{2k} + a_{2k+1}$ denote the even and odd numbered partial sums, respectively, then we have the following theorem, called the *alternating series test*:

Theorem. *The alternating series* $a_1 - a_2 + a_3 - \cdots + (-1)^{n+1} a_n + \cdots$ *converges to a sum* S *if* $a_1 \geq a_2 \geq a_3 \geq \cdots \geq 0$ *and* $\lim_{n \to \infty} a_n = 0$. *Furthermore, for every* k, $S_{2k} < S < S_{2k+1}$ *and for every* n, $|S - S_n| < a_{n+1}$.

For a visual proof, we use the hypotheses that $a_1 \geq a_2 \geq a_3 \geq \cdots \geq 0$ and $\lim_{n \to \infty} a_n = 0$ to construct the three columns of rectangles in Figure 39.1, placing the terms $a_1, a_2, a_3, \ldots$ on a vertical axis as a sequence of points decreasing to zero. Next we draw horizontal line segments one unit in length to bound a strip of gray rectangles with the indicated areas.

Let S be the area of the gray rectangles in Figure 39.1a, and S_{2k} and S_{2k+1} the areas of the gray rectangles in Figures 39.1b and c, respectively. It now follows that $S_{2k} < S < S_{2k+1}$.

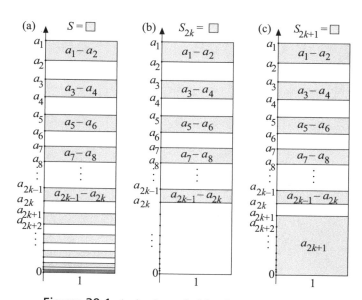

Figure 39.1. A visual proof of the alternating series test

The area of the gray rectangles in Figure 39.1a below the a_{2k+1} line segment represent the difference $S - S_{2k}$, and comparing Figures 39.1a and c shows that $|S - S_{2k}| < a_{2k+1}$. Similarly, the area of the white rectangles below the a_{2k+2} line segment in Figure 39.1a represent $S_{2k+1} - S$, and comparison with Figure 39.1c shows that $|S - S_{2k+1}| < a_{2k+2}$. Combining the last two inequalities yields $|S - S_n| < a_{n+1}$ for all n. Since $\lim_{n\to\infty} a_n = 0$ it now follows that $\lim_{n\to\infty} S_n = S$, so that series does indeed converge to S.

SOURCES: R. H. Hammack and D. W. Lyons, "Proof without words," *College Mathematics Journal*, **36** (2005), p. 72, and "Alternating series convergence: a visual proof," *Teaching Mathematics and Its Applications*, **25** (2006), pp. 58–60.

CAMEO 40

Approximating π with Maclaurin series

The Maclaurin series for the arctangent

$$\arctan x = x - \frac{x^3}{3} + \frac{x^5}{5} - \frac{x^7}{7} + \cdots = \sum_{n=0}^{\infty} (-1)^n \frac{x^{2n+1}}{2n+1}$$

converges for x in $[-1,1]$, and since $\arctan 1 = \pi/4$ we can use the series with $x = 1$ to approximate π. However, the convergence is much too slow to be practical. For example, the 1000th partial sum is only correct to two decimals. However, with the aid of *Hutton's formula*

$$\frac{\pi}{4} = 2\arctan\frac{1}{3} + \arctan\frac{1}{7}, \tag{40.1}$$

we can use the arctangent series to approximate π to as many decimal places as we wish rather easily.

Exercise 40.1. Prove Hutton's formula. While this can be done with trigonometric identities, it can also be done with Figure 40.1. (Hint: show that the acute angles in the lower left corner are $\arctan(1/3)$ for the two light gray right triangles and $\arctan(1/7)$ for the dark gray right triangle, and that the three angles sum to $\pi/4$.)

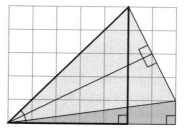

Figure 40.1. A visual proof of Hutton's formula

Using the arctangent series with $x = 1/3$ and $x = 1/7$ in Hutton's formula yields the following alternating series for π, where we have combined the series for $\arctan(1/3)$ and $\arctan(1/7)$ into a single series:

$$\pi = 4\sum_{n=0}^{\infty} \frac{(-1)^n}{2n+1}\left[2(1/3)^{2n+1} + (1/7)^{2n+1}\right].$$

Example 40.1. We now use this series to approximate π correct to eight decimal places. Since it's an alternating series, π lies between successive partial sums (which we denote by π_n). Computing the first few partial sums yields

$$\pi_0 = 3.238095238\cdots$$
$$\pi_1 = 3.135442536\cdots$$
$$\pi_2 = 3.142074498\cdots$$
$$\vdots$$
$$\pi_7 = 3.141592650\cdots$$
$$\pi_8 = 3.141592653\cdots$$

and we quit here since every number in the interval $(3.141592650\cdots, 3.141592653\cdots)$ has the same first eight decimals. Thus to eight decimal places, $\pi \approx 3.14159265$. Charles Hutton published (40.1) in 1776, and in 1789 Georg von Vega used it with the arctangent series to compute π to 143 decimal places, of which the first 126 were correct.

Exercise 40.2. A formula similar to Hutton's is *Strassnitzky's formula*:

$$\frac{\pi}{4} = \arctan\frac{1}{2} + \arctan\frac{1}{5} + \arctan\frac{1}{8}. \tag{40.2}$$

Prove Strassnitzky's formula (hint: see Figure 40.2), and use it to approximate π.

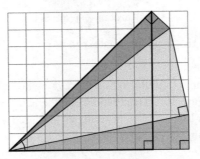

Figure 40.2. Strassnitzky's formula

L. K. Schulz von Strassnitzky provided (40.2) to Zacharias Dase in 1844, who then used it to compute π correct to 200 decimal places.

Prior to the work of Hutton and Strassnitzky, Isaac Newton wrote the book *Methodus Fluxionum et Serierum Infinitarum* in 1671, which contains an approximation to π correct to 16 decimal places based on what we now call the Maclaurin series for $\sqrt{1-x}$. We now recreate his approximation using modern terminology. It is based on Figure 40.3.

Exercise 40.3. Use Figure 40.3 to show that

$$\frac{\pi}{24} = \frac{\sqrt{3}}{32} + \int_0^{1/4} \sqrt{x-x^2}\, dx$$

and hence

$$\pi = \frac{3\sqrt{3}}{4} + 24\int_0^{1/4} \sqrt{x-x^2}\, dx. \tag{40.3}$$

CAMEO 40. Approximating π with Maclaurin series

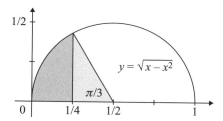

Figure 40.3. The semicircle in Newton's approximation of π

(Hint: the two shaded regions have a total area of $\pi/24$ (1/6 of the area $\pi/4$ of a circle of radius 1/2), while the light gray triangle has area $\sqrt{3}/32$.)

Exercise 40.4. Show that

$$\int_0^{1/4} \sqrt{x-x^2}\, dx = \frac{1}{12} - \frac{1}{5\cdot 2^5} - \frac{1}{4\cdot 7\cdot 2^7} - \frac{1\cdot 3}{4\cdot 6\cdot 9\cdot 2^9} - \frac{1\cdot 3\cdot 5}{4\cdot 6\cdot 8\cdot 11\cdot 2^{11}} - \cdots.$$

(Hint: write the integrand in (40.3) as $\sqrt{x}\sqrt{1-x}$, expand $\sqrt{1-x}$ into its Maclaurin series, multiply by $\sqrt{x}$ and integrate.)

The first nine terms of the series and (40.3) yield the approximation $\pi \approx 3.141592668\cdots$, which is accurate to seven decimal places.

Issac Newton

You may wonder how Newton computed $\sqrt{3}$ in his approximation. He used the Maclaurin series for $\sqrt{1-x}$ after writing $\sqrt{3}$ as $2\sqrt{1-(1/4)}$. At the time he wrote "I am ashamed to tell you to how many places of figures I carried these computations, having no other business at the time."

SOURCE: Newton's approximation of π is adapted from Chapter 7 in W. Dunham, *Journey Through Genius: The Great Theorems of Mathematics*, John Wiley & Sons, Inc., New York, 1990.

PART IV
Additional Topics

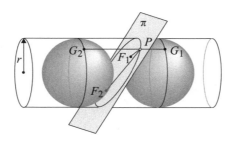

CAMEO 41

The hyperbolic functions I: Definitions

In most calculus texts hyperbolic functions are defined in terms of exponential functions: $\cosh u = (e^u + e^{-u})/2$ and $\sinh u = (e^u - e^{-u})/2$. Later certain identities are verified, and the source of the name "hyperbolic" is revealed: the points $(\cosh u, \sinh u)$ lie on the right-hand branch of the unit hyperbola $x^2 - y^2 = 1$. But this observation provides no *motivation* for the choice of these particular combinations of exponential functions in defining $\cosh u$ and $\sinh u$.

Recall that the circular functions are generally defined as coordinates of points on the unit circle $x^2 + y^2 = 1$. If θ represents the radian measure of the signed angle between the positive x-axis and the radius to a point P on the unit circle (by "signed" we mean that the angle is positive if the angle is measured in the counterclockwise direction, and negative in the clockwise direction), then the coordinates of P are defined to be $(\cos\theta, \sin\theta)$. This is equivalent to letting $\theta/2$ denote the signed area of the circular sector swept out by the radius OP, as in Figure 41.1a.

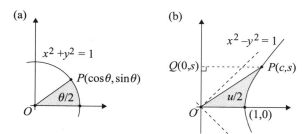

Figure 41.1. Defining the hyperbolic functions

Let's see what happens if we try to define the hyperbolic functions in an analogous manner by replacing the unit circle with the unit hyperbola. For a real number u, let a ray from the origin intersect the hyperbola at a point (c, s) so that the signed area of the hyperbolic sector is $u/2$, as illustrated in Figure 41.1b for $u > 0$ (the signed area of the sector is positive when it lies above the x-axis and negative when it lies below the x-axis). Then we define $\sinh u$ and $\cosh u$ to be the coordinates of the point on the hyperbola, that is, $(c, s) = (\cosh u, \sinh u)$.

We can now use an integral to express the area $u/2$ of the hyperbolic sector in terms of c and s, since it is the area to the left of the hyperbola for y between 0 and s minus the area of the right triangle OPQ:

$$\frac{u}{2} = \int_0^s \sqrt{y^2 + 1}\, dy - \frac{1}{2}cs.$$

121

CAMEO 41. The hyperbolic functions I: Definitions

The trigonometric substitution $y = \tan t$ followed by integration by parts yields

$$\int_0^s \sqrt{y^2+1}\, dy = \int_0^{\arctan s} \sec^3 t\, dt = \frac{1}{2}[\sec t \tan t + \ln|\sec t + \tan t|]_0^{\arctan s}.$$

Since $\tan(\arctan s) = s$ and $\sec(\arctan s) = \sqrt{s^2+1} = c$, we have

$$\int_0^s \sqrt{y^2+1}\, dy = \frac{1}{2}[cs + \ln|c+s|]$$

and thus

$$\frac{u}{2} = \frac{1}{2}[cs + \ln|c+s|] - \frac{1}{2}cs = \frac{1}{2}\ln(c+s),$$

where we have removed the absolute value bars in the logarithm since $c > s$. Hence $u = \ln(c+s)$, or equivalently

$$c + s = e^u. \tag{41.1}$$

But the point (c,s) lies on the hyperbola so that $c^2 - s^2 = 1$ and thus

$$c - s = \frac{c^2 - s^2}{c+s} = \frac{1}{e^u} = e^{-u}. \tag{41.2}$$

Solving (41.1) and (41.2) simultaneously yields the desired result:

$$\boxed{\,c = \cosh u = \frac{e^u + e^{-u}}{2} \quad \text{and} \quad s = \sinh u = \frac{e^u - e^{-u}}{2}.\,}$$

Exercise 41.1. Use integration in polar coordinates to find the area $u/2$ of the shaded region in Figure 41.1b. (Hint: since the polar equation of the hyperbola is $r^2 = \sec 2\theta$, integration by parts is not required.)

Exercise 41.2. Let h be a function given by $h(x) = f(x) + g(x)$ where f is an even function and g is an odd function. In this case, the functions f and g are called the *even and odd parts* of h. For example, if $h(x) = 1/(x^2 - 2x + 2)$, then $f(x) = (x^2 + 2)/(x^4 + 4)$ and $g(x) = 2x/(x^4 + 4)$. See Figure 41.2.

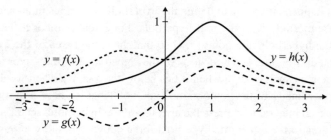

Figure 41.2. Graphs of the even and odd parts of a function

CAMEO 41. The hyperbolic functions I: Definitions

(a) Verify that the functions f and g sum to h.

(b) Show that if the domain of h is symmetric with respect to the origin (i.e., if a is in the domain so is $-a$), then there exists an even function f and an odd function g with the same domain such that $h(x) = f(x) + g(x)$ for all x in the domain of h. (Hint: if $h(x) = f(x) + g(x)$ with f even and g odd, what can you say about $h(-x)$?)

(c) Find the even and odd parts of the exponential function $h(x) = e^x$.

CAMEO 42

The hyperbolic functions II: Are they circular?

Admittedly the graphs of the hyperbolic sine and hyperbolic cosine look nothing like the graphs of the circular sine and cosine. But have you ever compared the graphs of all six hyperbolic functions to the graphs of all six circular functions? In Figure 42.1a we've graphed the six hyperbolic functions over the interval $[-2.25, 2.25]$, and in Figure 42.1b the six circular functions over the interval $(-\pi/2, \pi/2)$:

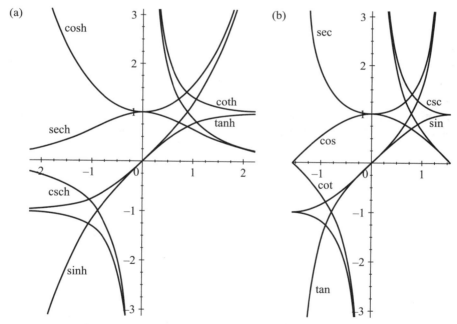

Figure 42.1. Graphs of the hyperbolic and circular functions

The similarity of the two sets of graphs is striking, which suggests that there may be functional relationships between the two sets of functions. To explore the relationships, recall that $(\cosh u, \sinh u)$ is a point on the right-hand branch of the unit hyperbola $x^2 - y^2 = 1$, draw a ray from the origin to the point $(1, \sinh u)$, and let ϕ denote the angle that the ray makes with the positive x-axis, as illustrated in Figure 42.2.

125

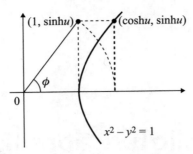

Figure 42.2. Relating the hyperbolic and circular functions

Now we evaluate the circular functions of ϕ, first noting that the length of the ray from the origin to the point $(1, \sinh u)$ is $\cosh u$. Consequently, we have

$$\tan\phi = \sinh u, \quad \sec\phi = \cosh u, \quad \cot\phi = \csch u, \tag{42.1}$$
$$\sin\phi = \tanh u, \quad \cos\phi = \sech u, \quad \csc\phi = \coth u,$$

and so each hyperbolic function (of u) is a circular function (of ϕ).

The angle ϕ can be expressed in six ways:

$$\phi = \arctan(\sinh u) = \arcsec(\cosh u) = \arcsin(\tanh u)$$
$$= \arccot(\csch u) = \arccos(\sech u) = \arccsc(\coth u).$$

The angle ϕ is called the *Gudermannian* of u, named for the German mathematician Christoph Gudermann (1798–1852) and written $\phi = \gd u$. Hence we have $\sinh u = \tan(\gd u)$, $\cosh u = \sec(\gd u)$, etc., explaining the appearance of the graphs in Figure 42.1.

The domain and range of $\phi = \gd u$ are $(-\infty, \infty)$ and $(-\pi/2, \pi/2)$, respectively. The Gudermannian is graphed in Figure 42.3 along with its horizontal asymptotes.

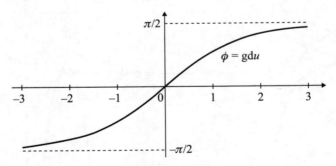

Figure 42.3. The graph of $\phi = \gd u$

Example 42.1. One of the hyperbolic Pythagorean identities, $1 + \sinh^2 u = \cosh^2 u$, follows from the definitions of the functions. The other two follow from the circular Pythagorean identities applied to the Gudermannian: for all real u,

$$\sin^2(\gd u) + \cos^2(\gd u) = 1 \text{ implies } \tanh^2 u + \sech^2 u = 1;$$
$$1 + \cot^2(\gd u) = \csc^2(\gd u) \text{ implies } 1 + \csch^2 u = \coth^2 u.$$

CAMEO 42. The hyperbolic functions II: Are they circular?

See Figure 42.4 for an illustration of the hyperbolic Pythagorean identities (angles marked ∡ have measure $\phi = \operatorname{gd} u$). Ratios of corresponding sides in similar triangles yields identities such as $\tanh u/1 = \sinh u/\cosh u$, $\coth u/1 = 1/\tanh u$, etc.

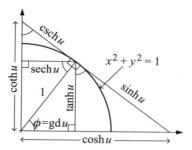

Figure 42.4. The hyperbolic Pythagorean identities

Example 42.2. The Gudermannian has a simple derivative:

$$\frac{d}{du}\operatorname{gd} u = \frac{d}{du}\arctan(\sinh u) = \frac{\cosh u}{1+\sinh^2 u} = \operatorname{sech} u,$$

which simplifies finding derivatives of the hyperbolic functions. For example,

$$\frac{d}{du}\sinh u = \frac{d}{du}\tan(\operatorname{gd} u) = \sec^2(\operatorname{gd} u)\operatorname{sech} u = \cosh^2 u \operatorname{sech} u = \cosh u.$$

Exercise 42.1. Use (42.1), the derivative of the Gudermannian, and the chain rule to evaluate the derivatives of the other five hyperbolic functions.

Example 42.2. Since the derivative of the Gudermannian $\phi = \operatorname{gd} u$ is positive on $(-\infty, \infty)$, $\phi = \operatorname{gd} u$ has a differentiable inverse $u = \operatorname{gd}^{-1}\phi$, and

$$\frac{du}{d\phi} = \frac{1}{d\phi/du} = \frac{1}{\operatorname{sech} u} = \cosh u = \sec\phi.$$

Hence

$$\operatorname{gd}^{-1}\phi = \int_0^\phi \sec t\, dt = \ln(\sec\phi + \tan\phi).$$

With the inverse of the Gudermannian, we can now find the inverse hyperbolic functions and their derivatives. For example, if $y = \sinh u = \tan(\operatorname{gd} u)$, then

$$u = \sinh^{-1} y = \operatorname{gd}^{-1}(\arctan y) = \ln(y + \sqrt{1+y^2})$$

and

$$\frac{d}{dy}\sinh^{-1} y = \frac{d}{dy}\operatorname{gd}^{-1}(\arctan y) = \frac{\sec(\arctan y)}{1+y^2} = \frac{1}{\sqrt{1+y^2}}, \text{ etc.}$$

Exercise 42.2. Establish the following properties of the Gudermannian:

a. $gd(-u) = -gd\,u$

b. $\lim_{u \to \infty} gd\,u = \pi/2$

c. $\lim_{u \to -\infty} gd\,u = -\pi/2$

d. $e^x = \sec(gd\,x) + \tan(gd\,x)$

e. $\tanh(\frac{1}{2}x) = \tan(\frac{1}{2}gd\,x)$

f. $gd\,x = 2\arctan e^x - \frac{\pi}{2}$.

SOURCE: J. M. H. Peters, "The Gudermannian," *Mathematical Gazette*, **68** (1984), pp. 192–196.

CAMEO 43

The conic sections

Many calculus texts define conic sections as, naturally, sections of a cone, and illustrate them as shown in Figure 43.1, where we see both nappes (a *nappe* of a cone is one of the two portions of a double cone) of three cones, and a parabola, an ellipse, a circle, and a hyperbola as conic sections.

Figure 43.1. The conic sections

However, to find either the rectangular or polar equations of a conic, a different definition is used, involving a fixed point (the *focus* of the conic) and a fixed line (the *directrix* of the conic). What is missing is a justification that the two definitions are equivalent, that is, the same curves result from either procedure.

The following theorem and elegant geometric proof of the equivalence of the conic-section and focus-directrix approaches is due to Germinal Pierre Dandelin (1794–1847) and Adolphe Quetelet (1796–1874). The following lemma is essential to their proof.

129

Germinal Pierre Dandelin and Adolphe Quetelet

Lemma. *The lengths of any two line segments from a point to a plane are inversely proportional to the sines of the angles that the line segments make with the plane.*

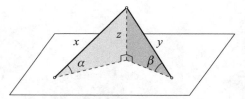

Figure 43.2. The proof of the lemma

See Figure 43.2, and observe that $z = x \sin\alpha = y \sin\beta$, hence $x/y = \sin\beta/\sin\alpha$.

Theorem. *Let π denote a plane that intersects a right circular cone in a conic section, and consider a sphere tangent to the cone and tangent to π at a point F (see Figure 43.3). Let π' denote the plane determined by the circle of tangency of the sphere and the cone, and let d denote the line of intersection of π and π'. Let P be any point on the conic section, and let D denote the foot of the line segment from P perpendicular to d. Then the ratio $|PF|/|PD|$ is a constant.*

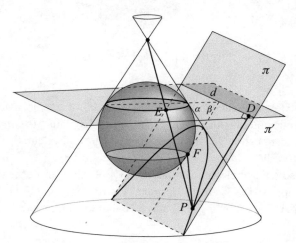

Figure 43.3. The proof of the theorem

CAMEO 43. The conic sections

To prove the theorem, we let E denote the point of intersection of the element of the cone (a line on the cone passing through the vertex) through P and the circle of tangency of the sphere. Then $|PF| = |PE|$, since two line segments from P tangent to a sphere must have the same length. Let α denote the angle that every element of the cone makes with π' and let β denote the angle between π and π'. Then $|PF|/|PD| = |PE|/|PD| = \sin\beta/\sin\alpha$, and $\sin\beta/\sin\alpha$ is a constant.

The point F in the proof is the *focus* of the conic section, and the line d the *directrix*. The constant $\sin\beta/\sin\alpha$ is often denoted by ε, the *eccentricity* of the conic section. When π is parallel to one and only one element of the cone, $\alpha = \beta$, $\varepsilon = 1$, and the conic is a parabola; when π cuts every element of one nappe, $\alpha > \beta$, $\varepsilon < 1$, and the conic is an ellipse; when π cuts both nappes of the cone, $\alpha < \beta$, $\varepsilon > 1$, and the conic is a hyperbola.

Ellipses (and circles) also result from the intersection of a cylinder with a plane that cuts all elements of the cylinder. When the plane is perpendicular to the axis of the cylinder the curve in the section of the cylinder is a circle. When the plane is not perpendicular to the axis of the cylinder, we can show that the curve in the section is an ellipse, that is, there are two points F_1 and F_2 (the *foci*, plural of focus) and a constant c such that for every point P on the curve we have $|PF_1| + |PF_2| = c$. See Figure 43.4.

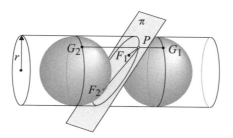

Figure 43.4. A cylindrical section

Consider two spheres of radius r inscribed in a cylinder of radius r and tangent to the intersecting plane π. Let F_1 be the point of tangency of the sphere on the right and F_2 the point of tangency of the sphere on the left. If P is any point on the curve, $|PF_1| = |PG_1|$, where G_1 is the intersection of the equator of the sphere on the right with the element of the cylinder that contains P. Analogously, $|PF_2| = |PG_2|$ so that

$$|PF_1| + |PF_2| = |PG_1| + |PG_2| = |G_1G_2| = c,$$

where c is the distance between the equators of the two spheres.

You may have noticed that we have two different descriptions for the ellipse, one with a cone (a single focus and the directrix) and one with the cylinder (two foci). But when the plane π passes through all the elements of the cone in Figure 43.3, we can inscribe a second sphere tangent to π at a point F'. Let the element of the cone through P and E intersect the circle of tangency of the second sphere at E'. Then $|PF'| = |PE'|$, which along with $|PF| = |PE|$ yields $|PF| + |PF'| = |PE| + |PE'|$. But $|PE| + |PE'|$ is a constant, the distance between the circles of tangency of the two spheres. So the curves in a plane intersecting all the elements of a cone and a plane intersecting all the elements of a cylinder are the same, the locus of points P such that the sum of the distances from two fixed points is a constant.

SOURCE: H. Eves, *An Introduction to the History of Mathematics, Fifth Edition*, Saunders College Publishing Company, Philadelphia, 1983.

CAMEO 44

The conic sections revisited

In the preceding Cameo we demonstrated the equivalence of the conic-section and the directrix-focus definitions of the conic sections. We now derive the familiar Cartesian equations of the conic sections from their definitions as intersections of a cone and a plane. Rather than using a fixed cone intersected by various planes, we rotate the axes and use a fixed plane.

For simplicity we consider a standard cone $z^2 = x^2 + y^2$ in the xyz-coordinate system, as illustrated in Figure 44.1, and rotate the y- and z-axes through an angle θ ($0 \le \theta \le \pi/2$) in the yz-plane to obtain an $x\bar{y}\bar{z}$-coordinate system. Note that for an observer in the space of the figure, the axes remain fixed while the cone rotates.

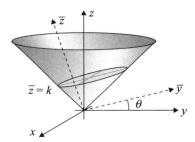

Figure 44.1. A standard cone with a rotation of axes

The y, z, $\bar{y}$, and $\bar{z}$ coordinates are related by $y = \bar{y} \cos\theta - \bar{z} \sin\theta$ and $z = \bar{y} \sin\theta + \bar{z} \cos\theta$ so that in the $x\bar{y}\bar{z}$-coordinate system the cone is given by

$$(\bar{y} \sin\theta + \bar{z} \cos\theta)^2 = x^2 + (\bar{y} \cos\theta - \bar{z} \sin\theta)^2.$$

To find the equation of the intersection of the plane $\bar{z} = k$ with the cone we set $\bar{z} = k$, expand, and simplify to get

$$x^2 + (\cos 2\theta)\bar{y}^2 - (2k \sin 2\theta)\bar{y} = k^2 \cos 2\theta. \tag{44.1}$$

If we replace y by $\bar{y}$ and set $\theta = \pi/4$ we have $x^2 - 2ky = 0$, which is a parabola, and when $\theta \ne \pi/4$ we can complete the square in (44.1) to obtain the standard form for the equation of a translated conic,

$$\frac{x^2}{\cos 2\theta} + \frac{(y - k\tan 2\theta)^2}{1} = k^2 \sec^2 2\theta.$$

We now consider four cases: (i) when $\theta = 0$ we obtain the circle $x^2 + y^2 = k^2$; (ii) when $0 < \theta < \pi/4$ we have an ellipse since $\cos 2\theta > 0$; (iii) when $\pi/4 < \theta < \pi/2$ we have a hyperbola since $\cos 2\theta < 0$; and (iv) when $\theta = \pi/2$ we have $x^2 - y^2 = k^2$, a hyperbola for $k \neq 0$ and two intersecting lines for $k = 0$. In each case, the value of $\cos 2\theta$ determines the nature of the conic.

Exercise 44.1. Use the same technique to find the equation for the intersection of a right circular cylinder and a plane. (Hint: In xyz-coordinates use the cylinder $x^2 + y^2 = r^2$, and after rotation of the y- and z-axes as above set $\bar{z} = 0$.)

SOURCE: M. R. Cullen, "Cylinder and cone cutting," *College Mathematics Journal*, **28** (1997), pp. 122–123.

CAMEO 45

The AM-GM inequality for n positive numbers

In Cameo 11 we encountered the arithmetic mean-geometric mean (AM-GM) inequality for two positive numbers, and in Cameo 15 the AM-GM inequality for three and for four positive numbers. The inequality actually holds for any finite number n of positive numbers $a_1, a_2, a_3, \ldots, a_n$.

Theorem. *The geometric mean G and the arithmetic mean A of n numbers $a_1, a_2, a_3, \ldots, a_n$ are given by*

$$G = \sqrt[n]{a_1 a_2 a_3 \ldots a_n} \quad \text{and} \quad A = \frac{a_1 + a_2 + a_3 + \cdots + a_n}{n}$$

and satisfy

$$\boxed{\sqrt[n]{a_1 a_2 a_3 \ldots a_n} \leq \frac{a_1 + a_2 + a_3 + \cdots + a_n}{n}} \qquad (45.1)$$

with equality if and only if $a_1 = a_2 = a_3 = \cdots = a_n$.

The proof begins with the simple inequality $ex \leq e^x$ (with equality if and only if $x = 1$). In Figure 45.1 we see that the line $y = ex$ is tangent to $y = e^x$ at $(1, e)$, and since the graph of $y = e^x$ in concave up, the inequality follows.

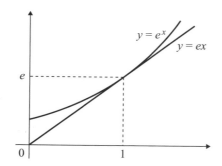

Figure 45.1. Graphs of $y = e^x$ and $y = ex$

With $x = a_i/G$ we have, for each i in $\{1, 2, 3, \ldots, n\}$,

$$e \frac{a_i}{G} \leq e^{a_i/G}$$

with equality if and only if $a_i = G$. Multiplying the n inequalities together yields

$$e^n \cdot \frac{a_1}{G} \cdot \frac{a_2}{G} \cdot \frac{a_3}{G} \cdots \frac{a_n}{G} \leq e^{(a_1+a_2+a_3+\cdots+a_n)/G}.$$

The inequality simplifies to $e^n \leq e^{nA/G}$, or $G \leq A$, with equality if and only if $a_1 = a_2 = a_3 = \cdots = a_n = G$.

Example 45.1. The arithmetic mean of the set $\{1, 2, 3, \ldots, n\}$ of the first n positive integers is $A_n = (n+1)/2$, while the geometric mean is $G_n = \sqrt[n]{n!}$. While both means increase without bound as $n \to \infty$, their ratio (which is greater than 1 for each $n \geq 2$) has a finite limit. Using (22.1) we have

$$\lim_{n \to \infty} \frac{A_n}{G_n} = \lim_{n \to \infty} \frac{n+1}{2\sqrt[n]{n!}} = \lim_{n \to \infty} \frac{n+1}{2n} \cdot \frac{n}{\sqrt[n]{n!}} = \frac{e}{2}.$$

Exercise 45.1. Show that $n! < [(n+1)/2]^n$ for $n \geq 2$.

Example 45.2. Let $a_1, a_2, \ldots, a_n$ and $b_1, b_2, \ldots, b_n$ be nonnegative real numbers. Show that

$$(a_1 a_2 \cdots a_n)^{1/n} + (b_1 b_2 \cdots b_n)^{1/n} \leq [(a_1+b_1)(a_2+b_2)\cdots(a_n+b_n)]^{1/n}.$$

(This is problem A2 from the 2003 Putnam Competition discussed in Cameo 20.)

If $a_i + b_i = 0$ for some i, then both sides of the inequality are 0, so assume $a_i + b_i > 0$ for every i. Then (45.1) yields

$$\left(\frac{a_1}{a_1+b_1} \cdot \frac{a_2}{a_2+b_2} \cdots \cdot \frac{a_n}{a_n+b_n}\right)^{1/n} \leq \frac{1}{n}\left(\frac{a_1}{a_1+b_1} + \frac{a_2}{a_2+b_2} + \cdots + \frac{a_n}{a_n+b_n}\right)$$

and

$$\left(\frac{b_1}{a_1+b_1} \cdot \frac{b_2}{a_2+b_2} \cdots \cdot \frac{b_n}{a_n+b_n}\right)^{1/n} \leq \frac{1}{n}\left(\frac{b_1}{a_1+b_1} + \frac{b_2}{a_2+b_2} + \cdots + \frac{b_n}{a_n+b_n}\right).$$

Adding the two inequalities yields

$$\left(\frac{a_1 a_2 \cdots a_n}{(a_1+b_1)(a_2+b_2)\cdots(a_n+b_n)}\right)^{1/n} + \left(\frac{b_1 b_2 \cdots b_n}{(a_1+b_1)(a_2+b_2)\cdots(a_n+b_n)}\right)^{1/n} \leq 1,$$

which is equivalent to the desired inequality.

Example 45.3. The geometric mean for n numbers can be used to explain the relationship between the ratio test and the root test for absolute convergence of an infinite series $\sum_{n=0}^{\infty} a_n$ with nonzero terms. The ratio test states that if $\lim_{n \to \infty} |a_{n+1}/a_n| < 1$ then the series converges absolutely, whereas the root test states that if $\lim_{n \to \infty} |a_n|^{1/n} < 1$ then the series converges absolutely. The connection between the two involves the consecutive ratios $|a_{n+1}/a_n|$ in the ratio test:

$$\lim_{n \to \infty} |a_n|^{1/n} = \lim_{n \to \infty} \left|\frac{a_n}{a_{n-1}} \cdot \frac{a_{n-1}}{a_{n-2}} \cdots \frac{a_2}{a_1} \cdot \frac{a_1}{a_0} \cdot a_0\right|^{1/n}$$

$$= \lim_{n \to \infty} \left(\left|\frac{a_n}{a_{n-1}}\right| \cdot \left|\frac{a_{n-1}}{a_{n-2}}\right| \cdots \left|\frac{a_2}{a_1}\right| \cdot \left|\frac{a_1}{a_0}\right|\right)^{1/n}$$

CAMEO 45. The AM-GM inequality for n positive numbers

(the last equality follows from $\lim_{n\to\infty} |a_0|^{1/n} = 1$). Thus the limit of the nth root of the nth term is the limit of the geometric mean of the first n consecutive ratios.

While the ratio test depends on the behavior (in the limit) of each consecutive ratio, the root test depends only on the average behavior (in the geometric mean sense) of the ratios. If all the ratios become small, then the geometric mean will become small; however, the converse is false, which is why the root test is stronger than the ratio test.

SOURCES: The proof of (45.1) is adapted from N. Schaumberger, "A coordinate approach to the AM-GM inequality," *Mathematics Magazine*, **64** (1991), p. 273, and Example 45.3 is from D. Cruz-Uribe, "The relation between the root and ratio tests," *Mathematics Magazine*, **70** (1997), pp. 214–215.

PART V
Appendix: Some Precalculus Topics

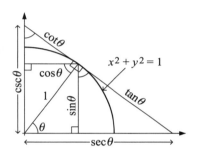

CAMEO 46

Are all parabolas similar?

In many calculus texts one finds examples and exercises concerning the properties of parabolas and their tangent lines. Here are a couple of examples:

1. Let P be a point on a parabola with focus F, and let Q be the point on the directrix closest to P. Show that the tangent line at P bisects angle FPQ.

2. Let P be a point on the parabola with focus F, and let R be the point where the tangent line at P intersects the tangent line at the vertex. Show that triangle FPR is a right triangle.

When the examples and exercises such as those above involve the focus and/or the directrix of the parabola, the texts often use $x^2 = 4py$ and $y^2 = 4px$ (or $y = kx^2$ and $x = ky^2$) as equations of general parabolas. Students often ask "Why can't we just use $y = x^2$, or $x = y^2$?" An answer is "Well, you could, *if* all parabolas were similar." At first glance parabolas seem to have different shapes, as can be seen by graphing three different parabolas, such as $y = x^2$, $y = 4x^2$, and $y = (1/4)x^2$ in the same window (e.g., the standard one $[-10, 10] \times [-10, 10]$) on a graphing calculator (see Figure 46.1 for a computer generated version).

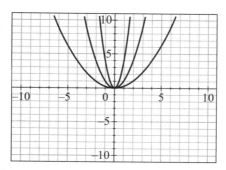

Figure 46.1. Graphs of the three parabolas in $[-10, 10] \times [-10, 10]$

But parabolas, like circles, *are* all similar, as a simple geometric exploration reveals. Ask the students to look at the graphs of the three parabolas in different square windows (recall that square windows look rectangular on most calculator screens):

a) graph $y = x^2$ in $[-2, 2] \times [0, 4]$

b) graph $y = 4x^2$ in $[-1/2, 1/2] \times [0, 1]$

c) graph $y = (1/4)x^2$ in $[-8, 8] \times [0, 16]$.

141

The students will see exactly the same shape in each graph, with the parabola passing through the top left and top right vertices of the window and the midpoint of the bottom of the window, as seen in Figure 46.2 (again computer generated).

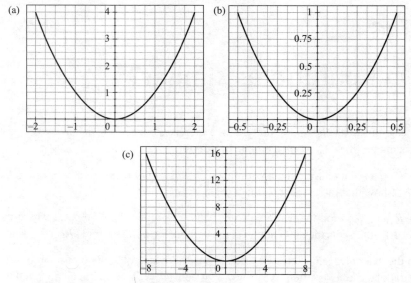

Figure 46.2. Graphs of (a) $y = x^2$, (b) $y = 4x^2$, and (c) $y = (1/4)x^2$

Thus zooming in by a factor of 4 (from $[-2, 2] \times [0, 4]$ to $[-1/2, 1/2] \times [0, 1]$) makes the graph of $y = 4x^2$ look just like the graph of $y = x^2$; and zooming out by a factor of 4 (from $[-2, 2] \times [0, 4]$ to $[-8, 8] \times [0, 16]$) makes the graph of $y = (1/4)x^2$ look just like the graph of $y = x^2$.

Now it's a simple matter to show algebraically that $y = kx^2$ ($k \neq 0$) is similar to $y = x^2$: Multiply both sides of $y = kx^2$ by k to yield $ky = (kx)^2$ and let $\bar{y} = ky$ and $\bar{x} = kx$ (zooming in when $k > 0$, zooming out when $k < 0$).

Exercise 46.1. Solve the two problems at the beginning of this Cameo. (Hint: solve the two problems simultaneously by showing that $\triangle FPQ$ is isosceles, FQ passes through R, and $PR \perp FQ$. See Figure 46.3.)

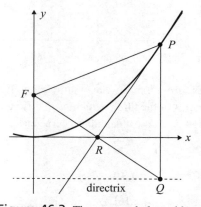

Figure 46.3. The two parabola problems

CAMEO 47

Basic trigonometric identities

Familiarity with basic trigonometric identities—the three Pythagorean relations as well as all the reciprocal and quotient identities—is essential for success in using the trigonometric substitution technique for integration. All the identities can be found in Figure 47.1 (for the case of first quadrant angles) by using the Pythagorean theorem and properties of similar right triangles. It can be a profitable exercise to find all of them in the Figure. (Hint: each angle marked ∡ has measure θ.)

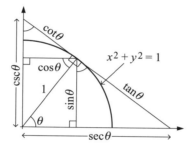

Figure 47.1. Trigonometric identities

For example, using the Pythagorean theorem on right triangles with an acute angle at the origin yields the three Pythagorean identities ($\sin^2\theta + \cos^2\theta = 1$, $\tan^2\theta + 1 = \sec^2\theta$, and $\cot^2\theta + 1 = \csc^2\theta$) while ratios of corresponding sides in two of these triangles yields identities such as $\tan\theta/1 = \sin\theta/\cos\theta$, $\cot\theta/1 = 1/\tan\theta$, etc.

Exercise 47.1. Use Figure 47.1 to establish these less-familiar identities:

(a) $(\tan\theta + \cot\theta)^2 = \sec^2\theta + \csc^2\theta$ (b) $\dfrac{\sec\theta - \cos\theta}{\cos\theta} = \dfrac{\sin\theta}{\csc\theta - \sin\theta}.$

SOURCE: R. D. Carmichael, "On the representation of the trigonometric functions by lines," *American Mathematical Monthly,* **15** (1908), pp. 199–200.

CAMEO 48

The addition formulas for the sine and cosine

The addition formulas are used in simplifying the difference quotient when using the definition of the derivative to differentiate the sine and cosine. We use Figure 48.1 to illustrate both formulas for acute angles α and β whose sum is less than $\pi/2$.

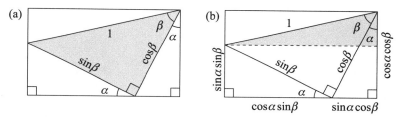

Figure 48.1. Illustrating the addition formulas for the sine and cosine

If we let the sides of the gray triangle in Figure 48.1a be $\sin\beta$, $\cos\beta$, and 1, then it is easy to compute the lengths of the sides of the two triangles with acute angle α, as shown in Figure 48.1b. Evaluating the lengths of the legs of the gray right triangle in Figure 48.1b yields

$$\sin(\alpha + \beta) = \sin\alpha \cos\beta + \cos\alpha \sin\beta,$$
$$\cos(\alpha + \beta) = \cos\alpha \cos\beta - \sin\alpha \sin\beta.$$

Exercise 48.1. Use Figure 48.2 to illustrate the subtraction formulas:

$$\sin(\alpha - \beta) = \sin\alpha \cos\beta - \cos\alpha \sin\beta,$$
$$\cos(\alpha - \beta) = \cos\alpha \cos\beta + \sin\alpha \sin\beta.$$

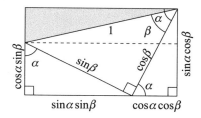

Figure 48.2. Illustrating the subtraction formulas for the sine and cosine

Exercise 48.2. Create figures to illustrate the addition and subtraction formulas for the tangent:

$$\tan(\alpha + \beta) = \frac{\tan\alpha + \tan\beta}{1 - \tan\alpha \tan\beta} \quad \text{and} \quad \tan(\alpha - \beta) = \frac{\tan\alpha - \tan\beta}{1 + \tan\alpha \tan\beta}.$$

SOURCE: RBN, "One figure, six identities," *College Mathematics Journal*, **31** (2000), pp. 145–146.

CAMEO 49

The double angle formulas

The double angle formulas for the sine and cosine are used in calculus for integrals of the form $\int \sin^{2m} x \cos^{2n} x\, dx$ where both m and n are nonnegative integers. The formulas are

$$\sin 2x = 2\sin x \cos x \quad \text{and} \quad \cos 2x = \begin{cases} 2\cos^2 x - 1 \\ 1 - 2\sin^2 x \\ \cos^2 x - \sin^2 x. \end{cases}$$

These are special cases of the addition formulas for the sine and cosine in Cameo 48, however they can also be illustrated with Figure 49.1, in which we inscribe a right triangle in a semicircle of radius 1. The angle at the origin marked ∡ has the value $2x$ since it is an exterior angle of an isosceles triangle with base angles each equal to x.

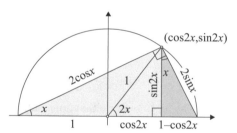

Figure 49.1. The double angle formulas

The large right triangle in two shades of gray has hypotenuse 2 and legs $2\sin x$ and $2\cos x$. In the light gray right triangle, we have $\sin x = \sin 2x/2\cos x$, hence $\sin 2x = 2\sin x \cos x$. In the light gray right triangle we also have $\cos x = (1 + \cos 2x)/2\cos x$ or $\cos 2x = 2\cos^2 x - 1$. In the dark gray right triangle, we have $\sin x = (1 - \cos 2x)/2\sin x$, so that $\cos 2x = 1 - 2\sin^2 x$. The Pythagorean relation yields the third form of the cosine formula.

For an alternate illustration of the double angle formulas (with $\cos 2x = \cos^2 x - \sin^2 x$), see Figure 49.2.

Closely related to the double angle formulas are the half angle formulas

$$\sin \frac{x}{2} = \pm\sqrt{\frac{1 - \cos x}{2}} \quad \text{and} \quad \cos \frac{x}{2} = \pm\sqrt{\frac{1 + \cos x}{2}}.$$

Exercise 49.1. Derive the half-angle formulas. When is the $+$ or $-$ sign used?

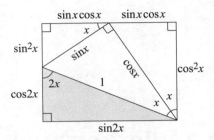

Figure 49.2. The double angle formulas, again

Exercise 49.2. Should you ever need them (you probably won't in calculus), there are *triple angle formulas*:

$$\sin 3x = 3\sin x - 4\sin^3 x \quad \text{and} \quad \cos 3x = 4\cos^3 x - 3\cos x.$$

Use Figure 49.3 and the double angle cosine formulas to derive them.

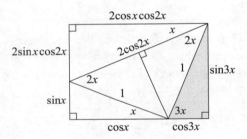

Figure 49.3. The triple angle formulas

SOURCES:

Figure 49.1 is from R. Woods, "The trigonometric functions of half or double an angle," *American Mathematical Monthly*, **43** (1936), pp. 174–175.

Figure 49.3 is from C. Alsina and RBN, "Proof without words: The triple angle sine and cosine formulas," *Mathematics Magazine*, **85** (2012), p. 43.

CAMEO 50

Completing the square

In calculus the completing the square technique is frequently used in the evaluation of integrals whose integrand involves a power of a quadratic expression with a nonzero linear term, usually encountered when studying trigonometric substitutions. Students may well have encountered the completing the square technique in elementary algebra, as it is the foundation for the quadratic formula.

The technique is based on the following identity: for any two real numbers x and a,

$$x^2 + 2ax = (x^2 + 2ax + a^2) - a^2 = (x+a)^2 - a^2. \tag{50.1}$$

Here is a geometric illustration of (50.1) for positive x and a:

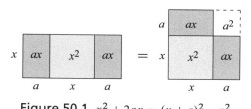

Figure 50.1. $x^2 + 2ax = (x+a)^2 - a^2$

Identity (50.1) can be used in two ways to complete the square in the general quadratic expression $ax^2 + bx + c$ ($a \neq 0$):

1. $ax^2 + bx + c = a(x^2 + \frac{b}{a}x) + c = a[(x + \frac{b}{2a})^2 - (\frac{b}{2a})^2] + c$.
2. $ax^2 + bx + c = \frac{1}{4a}(4a^2x^2 + 4abx) + c = \frac{1}{4a}[(2ax + b)^2 - b^2] + c$.

Completing the square can also be used to factor certain sums of squares by expressing the sum as a difference of squares. In Figure 50.2 we illustrate with $x^4 + 4a^4$:

Thus

$$\begin{aligned} x^4 + 4a^4 &= (x^4 + 4a^2x^2 + 4a^4) - 4a^2x^2 = (x^2 + 2a^2)^2 - (2ax)^2 \\ &= (x^2 + 2ax + 2a^2)(x^2 - 2ax + 2a^2). \end{aligned} \tag{50.2}$$

Combining (50.2) with (50.1) yields

$$x^4 + 4a^4 = [(x+a)^2 + a^2][(x-a)^2 + a^2].$$

Observe that in (50.1) we added and subtracted the missing constant in a squared polynomial, whereas in (50.2) we added and subtracted the missing middle term in a squared polynomial.

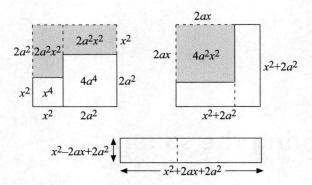

Figure 50.2. Completing another square

Example 50.1. The form of completing the square in (50.2) is the first step in the partial fraction decomposition of the integrand in $\int [1/(x^4+4)]dx$:

$$\frac{1}{x^4+4} = \frac{1}{(x^2+2x+2)(x^2-2x+2)}.$$

Exercise 50.1. Factor (a) x^6+1 and (b) x^8+x^4+1 into products of quadratic polynomials with real coefficients.

Solutions to the Exercises

Part I Limits and Differentiation

1.1. $\lim_{t\to 0} \frac{\tan t}{t} = \lim_{t\to 0} \frac{\sin t}{t} / \lim_{t\to 0} \cos t = 1/1 = 1$.

1.2. Since $\cos t < 1$ and $2 - 2\cos t < t^2$ for t in $(-\pi/2, 0) \cup (0, \pi/2)$, we have $0 < 1 - \cos t < t^2/2$. Division by t yields $0 < (1 - \cos t)/t < t/2$ for $t > 0$ and $0 > (1 - \cos t)/t > t/2$ for $t < 0$. Hence the two one-sided limits as $t \to 0$ are 0, and $\lim_{t\to 0} (1 - \cos t)/t = 0$.

2.1. The circumference of a regular n-gon inscribed in the unit circle is $2n \sin(\pi/n)$.

2.2. The sequence results from dividing the fourth column of Table 2.1 by the third column. The area of a regular n-gon circumscribed about the unit circle is $n \tan(\pi/n)$.

4.1. The function f is continuous since it is differentiable.

4.2. $\frac{d}{dx} f(x)g(x) = \frac{1}{2}(\frac{d}{dx}[f(x)+g(x)]^2 - \frac{d}{dx}[f(x)]^2 - \frac{d}{dx}[g(x)]^2)$
$= \frac{1}{2}(2[f(x)+g(x)][f'(x)+g'(x)] - 2f(x)f'(x) - 2g(x)g'(x))$
$= f(x)g'(x) + g(x)f'(x)$.

5.1. The function g is continuous since it is differentiable.

5.2. $\frac{d}{dt}\left(\frac{f(t)}{g(t)}\right) = f'(t) \cdot \frac{1}{g(t)} + f(t) \cdot \frac{-g'(t)}{[g(t)]^2} = \frac{g(t)f'(t) - f(t)g'(t)}{[g(t)]^2}$.

5.3. In the "proof" it is assumed that y is differentiable.

7.1. In Figure 7.1 let $x_0 = \cos\theta$. Since $\Delta\theta > 0$ we have $\Delta x < 0$, so that the length of the shorter leg of the dark gray triangle is $-\Delta x$. Hence

$$\frac{dx}{d\theta} \approx \frac{-\Delta x}{\Delta \theta} = \frac{-y_0}{1} = -\cos\phi = -\sin\theta.$$

8.1. Assume θ, $\Delta\theta$, and $\theta + \Delta\theta$ are in $(0, \pi/2)$, as illustrated in Figure 8.2. Then Area(sector OAB) $\leq$ Area ($\triangle OAD$) $\leq$ Area(sector OCD), so that

$$\frac{\Delta\theta \sec^2\theta}{2} \leq \frac{\tan(\theta + \Delta\theta) - \tan(\theta)}{2} \leq \frac{\Delta\theta \sec^2(\theta + \Delta\theta)}{2}.$$

Hence

$$\sec^2\theta \leq \frac{\tan(\theta + \Delta\theta) - \tan(\theta)}{\Delta\theta} \leq \sec^2(\theta + \Delta\theta)$$

and the result follows (since the secant, as the reciprocal of the cosine, is continuous on $(0, \pi/2)$).

10.1. The MVT yields $\frac{\ln b - \ln a}{b-a} = \frac{1}{c}$, so $c = \frac{b-a}{\ln b - \ln a}$.

10.2. The MVT yields $\frac{b \ln b - a \ln a}{b-a} = 1 + \ln c$, so $c = (1/e) \cdot (b^b/a^a)^{1/(b-a)}$.

11.1. The sum of the areas of the two triangles is at least the area of the rectangle, so $(\sqrt{a})^2/2 + (\sqrt{b})^2/2 \geq \sqrt{a}\sqrt{b}$, equivalent to (11.2).

11.2. $0 \leq (\sqrt{a} - \sqrt{b})^2 = a - 2\sqrt{ab} + b$, equivalent to (11.2).

11.3. (i) $\frac{\sqrt{a}}{\sqrt{b}} + \frac{\sqrt{b}}{\sqrt{a}} \geq 2$, hence $\frac{a+b}{2} \geq \sqrt{ab}$.

(ii) $\frac{x+(1/x)}{2} \geq \sqrt{x \cdot (1/x)} = 1$, hence $x + \frac{1}{x} \geq 2$.

11.4. The minimum value is 6. Using the hint, the expression becomes

$$\frac{a^2 - b^2}{a+b} = a - b = 3\left(x + \frac{1}{x}\right) \geq 3(2) = 6$$

with equality at $x = 1$, using either (11.1) or (11.2).

11.5. In each case, $y = 1 + rx$ is tangent to $y = (1+x)^r$ at $(0, 1)$. When $r > 1$ or $r < 0$, $y = (1+x)^r$ is concave up and $(1+x)^r \geq 1 + rx$. When $0 < r < 1$, $y = (1+x)^r$ is concave down and $(1+x)^r \leq 1 + rx$.

11.6. Using the hint, (11.4) becomes $(a/b)^r \leq 1 + (ra/b) - r$, and hence $a^r b^{1-r} \leq (1-r)b + ra$.

11.7. $y = x - 1$ is tangent to both $y = \ln x$ and $y = x \ln x$ at $(1, 0)$, while $y = \ln x$ is concave down and $y = x \ln x$ is concave up.

11.8. The inequality in Exercise 11.6 becomes $n \ln x^{1/n} \leq n(x^{1/n} - 1) \leq nx^{1/n} \ln x^{1/n}$, which simplifies to $\ln x \leq n(x^{1/n} - 1) \leq x^{1/n} \ln x$.

11.9. Taking the limit as $n \to \infty$ yields $\ln x \leq \lim_{n \to \infty} n(x^{1/n} - 1) \leq \ln x$ and hence $\lim_{n \to \infty} n(x^{1/n} - 1) = \ln x$ by the squeeze theorem.

11.10. (i) $y = x$ is tangent to both $y = \sin x$ and $y = \tan x$ at $(0, 0)$, while $y = \sin x$ is concave down and $y = \tan x$ is concave up, which establishes $\sin x < x < \tan x$. But $x < \tan x$ is equivalent to $x \cos x < \sin x$, hence $x \cos x < \sin x < x$. The inequalities are strict since x is in $(0, \pi/2)$.

(ii) Using the hint we have $\frac{d}{dx}\left(\frac{\sin x}{x}\right) = \frac{x \cos x - \sin x}{x^2} < 0$ and $\frac{d}{dx}\left(\frac{\tan x}{x}\right) = \frac{x \sec^2 x - \tan x}{x^2} = \frac{2x - \sin 2x}{2x^2 \cos^2 x} > 0$, so $(\sin x)/x$ is decreasing and $(\tan x)/x$ is increasing on $(0, \pi/2)$. Hence $\frac{\sin \beta}{\beta} > \frac{\sin \alpha}{\alpha}$ and $\frac{\tan \beta}{\beta} < \frac{\tan \alpha}{\alpha}$ on $(0, \pi/2)$.

14.1. No, since $dA/ds = 2s \neq 4s = P$ and $dV/ds = 3s^2 \neq 6s^2 = S$.

14.2. $A = (2x)^2 = 4x^2$ and $P = 4(2x) = 8x$ so that $dA/dx = P$, $V = (2x)^3 = 8x^3$, and $S = 6(2x)^2 = 24x^2$ so that $dV/dx = S$.

14.3. (a) $x = \frac{2A}{P} = \frac{s^2\sqrt{3}/2}{3s} = \frac{s\sqrt{3}}{6} = \frac{h}{3}$.

(b) $s = 2x\sqrt{3}$ so $P = 6x\sqrt{3}$, $A = 3x^2\sqrt{3}$, and $dA/dx = P$.

Solutions to the Exercises   153

14.4. Let $x = ts$. Then $V = c(x/t)^3$ and $S = k(x/t)^2$, so $dV/dx = S$ if and only if $t = 3c/k$. Thus $x = 3cs/k$. But $V/S = cx/kt = cs/k$, so $x = 3V/S$.

14.5. (a) If $R = mr$, $V = 2m\pi^2 r^3$ and $S = 4m\pi^2 r^2$, so $dV/dr = 6m\pi^2 r^2 \neq S$.
(b) $x = 3V/S = 6m\pi^2 r^3/4m\pi^2 r^2 = 3r/2$.

15.1. (i) Assume $P = 2(x + y)$ is constant, so that $y = P/2 - x$. Then $A(x) = xP/2 - x^2$ for $x \in [0, P/2]$, and $A'(x) = P/2 - 2x$ yields the critical point $x = P/4$. But $A(0) = 0 = A(P/2)$ and $A(P/4) > 0$, so the area is maximized at $x = y = P/4$. (ii) Assume $A = xy$ is constant so that $y = A/x$. Then $P(x) = 2(x + A/x)$ for $x > 0$, and $P'(x) = 2(1 - A/x^2)$ yields the critical point $x = \sqrt{A}$. But $P''(x) > 0$ for all $x > 0$, so we have an absolute minimum perimeter when $x = y = \sqrt{A}$.

15.2. Using the notation of Example 15.1, let $f(x) = \cot\theta = [x + (ab/x)]/(a - b)$ for $x > 0$. Then $f'(x) = [1 - (ab/x^2)]/(a - b)$ so that $x = \sqrt{ab}$ is the only critical point. Since $f''(x) > 0$ for all $x > 0$ $x = \sqrt{ab}$ minimizes $\cot\theta$, and hence maximizes θ.

15.3. Algebra easily establishes the identity. Since $x^2 + y^2 \geq 2xy$, etc., we have

$$x^2 + y^2 + z^2 - xy - yz - xz$$
$$= \frac{1}{2}[(x^2 + y^2 - 2xy) + (y^2 + z^2 - 2yz) + (x^2 + z^2 - 2xz)] \geq 0,$$

and hence $x^3 + y^3 + z^2 \geq 3xyz$.

15.4. Assume $V = \pi r^2 h/3$ is constant, so that $h = 3V/\pi r^2$. Let $f(r) = S^2 = \pi^2 r^2(r^2 + h^2) = \pi^2 r^4 + 9V^2/r^2$ for $r > 0$. Then $f'(r) = 4\pi^2 r^3 - 18V^2/r^3$ so that $r = (V/\pi)^{1/3} \sqrt[6]{9/2}$. Since $f''(r) > 0$ for all $r > 0$, this critical point yields an absolute minimum with $h/r = 3V/\pi r^3 = \sqrt{2}$.

15.5. Since V is fixed, $h = V/\pi r^2$ and $f(r) = C = (p + m)\pi r^2 + 2cV/r$ for $r > 0$. Hence $f'(r) = 2(p + m)\pi r - 2cV/r^2$, so that $r = \sqrt[3]{cV/(p + m)\pi}$ is the critical point. Since $f''(r) > 0$ this critical point yields an absolute minimum cost with $h/r = V/\pi r^3 = (p + m)/c$.

15.6. $S/6 = (ab + bc + ac)/3 \geq \sqrt[3]{ab \cdot bc \cdot ac} = V^{2/3}$, so $S \geq 6V^{2/3}$. Then

$$E^2/8 = (a^2 + b^2) + (b^2 + c^2) + (a^2 + c^2) + 4(ab + bc + ac)$$
$$\geq 6(ab + bc + ac) = 3S,$$

and hence $E^2/24 \geq S$, with equality if and only if $a = b = c$.

15.7. $\frac{a+b+c+d}{4} = \frac{1}{2}(\frac{a+b}{2} + \frac{c+d}{2}) \geq \frac{1}{2}(\sqrt{ab} + \sqrt{cd}) \geq \sqrt{\sqrt{ab}\sqrt{cd}} = \sqrt[4]{abcd}$.

Part II Integration

17.1. $\sum_{i=1}^{n} i + \sum_{i=1}^{n}(i - 1) = n^2 - 0$, so $\sum_{i=1}^{n}(2i - 1) = n^2$.

17.2. $\sum_{i=1}^{n} \frac{i(i + 1)}{2} + \sum_{i=1}^{n}(i + 1)i = (n + 2)\frac{n(n + 1)}{2}$, so $\sum_{i=1}^{n} \frac{i(i + 1)}{2} = n(n + 1)(n + 2)/6$.

17.3. $\sum_{i=1}^{n} \frac{i(i+1)(2i+1)}{6} \cdot 6i + \sum_{i=1}^{n} 3i(i-1) \cdot i^2 = 3n(n+1)\frac{n(n+1)(n+2)}{6}$, so

$\sum_{i=1}^{n} (5i^4 + i^2) = 3n(n+1)\frac{n(n+1)(n+2)}{6}$ and hence

$\sum_{i=1}^{n} i^4 = \frac{3n(n+1)-1}{5} \cdot \frac{n(n+1)(n+2)}{6} = \frac{n(n+1)(n+2)(3n^2+3n-1)}{30}$.

17.4. $\sum_{i=1}^{n} (i+3) \cdot \frac{3}{i(i+1)(i+2)(i+3)} - \sum_{i=1}^{n} \frac{1}{i(i+1)(i+2)} \cdot 1 =$

$\frac{-1}{(n+1)(n+2)(n+3)}(n+3) + 3 \cdot \frac{1}{6}$,

so $\sum_{i=1}^{n} \frac{1}{i(i+1)(i+2)} = \frac{1}{4} - \frac{1}{2(n+1)(n+2)}$.

17.5. $-\sum_{i=1}^{n} i \left(\frac{1}{2}\right)^i + \sum_{i=1}^{n} \left(\frac{1}{2}\right)^{i-1} \cdot 1 = n \left(\frac{1}{2}\right)^n$, so $\frac{1}{2}\sum_{i=1}^{n} i \left(\frac{1}{2}\right)^{i-1} =$

$-n\left(\frac{1}{2}\right)^n + \frac{1-(1/2)^n}{1-(1/2)}$, and hence $\sum_{i=1}^{n} i \left(\frac{1}{2}\right)^{i-1} = 4 - (n+2)\left(\frac{1}{2}\right)^{n-1}$.

18.1. From the hint we have $A_0 + A_1 = \int_p^q v\,du$, $A_0 + A_2 = \int_s^r u\,dv$, $A_1 + A_3 = qs$, $A_2 + A_3 = pr$, and (18.1) follows.

19.1. $|AC| = 2\sin(x/2) = 2z/\sqrt{1+z^2}$. Ratios of sides in similar triangles yields

$$\frac{\sin x}{2z/\sqrt{1+z^2}} = \frac{1}{\sqrt{1+z^2}}, \text{ hence } \sin x = \frac{2z}{1+z^2}$$

and

$$\frac{1-\cos x}{2z/\sqrt{1+z^2}} = \frac{z}{\sqrt{1+z^2}}, \text{ hence } \cos x = \frac{1-z^2}{1+z^2}.$$

19.2. The lengths of the hypotenuses of the three right triangles are (from shortest to longest) $z\sqrt{1+z^2}$, $\sqrt{1+z^2}$, and $1+z^2$. In the shaded right triangle we then have $\sin x = 2z/(1+z^2)$ and $\cos x = (1-z^2)/(1+z^2)$.

20.1. The trapezoidal rule is exact since setting x equal to a or b in (20.2) yields $\frac{1}{2}[f(a)+f(b)] = f(\frac{a+b}{2})$. Simpson's rule is exact since it is a weighted arithmetic mean of the midpoint and trapezoidal rules.

20.2. $\frac{1}{2}\left[\frac{1}{4+2^x} + \frac{1}{4+2^{4-x}}\right] = \frac{1}{2}\left[\frac{1}{4+2^x} + \frac{2^{x-2}}{2^x+4}\right] = \frac{1}{2}[\frac{1}{4}] = \frac{1}{8} = \frac{1}{4+2^2}$, hence $\int_0^4 \frac{dx}{4+2^x} = 4 \cdot \frac{1}{8} = \frac{1}{2}$.

20.3. $\frac{1}{2}[\arctan e^x + \arctan e^{-x}] = \frac{1}{2}[\frac{\pi}{2}] = \frac{\pi}{4} = \arctan e^0$, hence $\int_{-1}^{1} \arctan(e^x)dx = 2(\pi/4) = \pi/2$.

20.4. $\frac{1}{2}[\arccos(x^3) + \arccos(-x^3)] = \frac{1}{2}\arccos(-1) = \frac{1}{2}\pi = \arccos 0$, hence $\int_{-1}^{1} \arccos(x^3)dx = 2[\pi/2] = \pi$.

Solutions to the Exercises 155

20.5. $\frac{1}{2}[\frac{1}{x+\sqrt{x^2-2x+2}} + \frac{1}{2-x+\sqrt{x^2-2x+2}}] = \frac{1}{2}[\frac{2+2\sqrt{x^2-2x+2}}{2+2\sqrt{x^2-2x+2}}] = \frac{1}{2} = \frac{1}{1+\sqrt{1}}$,

hence $\int_0^2 \frac{dx}{x+\sqrt{x^2-2x+2}} = 2 \cdot \frac{1}{2} = 1$.

20.6. $\frac{1}{2}[\frac{1}{1+(\tan x)^{\sqrt{2}}} + \frac{dt}{1+[\tan(\pi/2-x)]^{\sqrt{2}}}] = \frac{1}{2}[\frac{1}{1+(\tan x)^{\sqrt{2}}} + \frac{1}{1+(\cot x)^{\sqrt{2}}}] =$

$\frac{1}{2}[\frac{1}{1+(\tan x)^{\sqrt{2}}} + \frac{(\tan x)^{\sqrt{2}}}{1+(\tan x)^{\sqrt{2}}}] = \frac{1}{2} = \frac{1}{1+(\tan(\pi/4))^{\sqrt{2}}}$, hence

$\int_0^{\pi/2} \frac{dx}{1+(\tan x)^{\sqrt{2}}} = \frac{\pi}{2} \cdot \frac{1}{2} = \frac{\pi}{4}$.

20.7. $\frac{1}{2}[\frac{\sqrt{\ln(9-x)}}{\sqrt{\ln(9-x)}+\sqrt{\ln(x+3)}} + \frac{\sqrt{\ln(x+3)}}{\sqrt{\ln(9-x)}+\sqrt{\ln(x+3)}}] = \frac{1}{2} = \frac{\sqrt{\ln 6}}{2\sqrt{\ln 6}}$,

hence $\int_2^4 \frac{\sqrt{\ln(9-x)}\,dx}{\sqrt{\ln(9-x)}+\sqrt{\ln(x+3)}} = 2 \cdot \frac{1}{2} = 1$.

21.1. Multiplying (21.2) by $n+1$ yields $1 < (n+1)\ln(1+1/n) < (n+1)/n$, and (21.4) follows upon exponentiation. The squeeze theorem now yields the desired limit.

21.2. Combine the second inequality in (21.3) with the first inequality in (21.4).

21.3. When $x > 1$ setting $(a,b) = (1,x)$ in (21.1) yields $\frac{1}{x} < \frac{\ln x}{x-1} < 1$ and when $0 < x < 1$ setting $(a,b) = (x,1)$ in (21.1) yields $1 < \frac{\ln x}{x-1} < \frac{1}{x}$. Applying the squeeze theorem yields $\lim_{x \to 1^+} \frac{\ln x}{x-1} = 1 = \lim_{x \to 1^-} \frac{\ln x}{x-1}$ so $\lim_{x \to 1} \frac{\ln x}{x-1} = 1$.

21.4. When $x > 0$ we have $\frac{1}{1+x} < \frac{\ln(x+1)}{x} < 1$, or $\frac{x}{x+1} < \ln(x+1) < x$. When $-1 < x < 0$ we have $1 < \frac{\ln(x+1)}{x} < \frac{1}{1+x}$, or $x > \ln(x+1) > \frac{x}{x+1}$. Hence $\frac{x}{x+1} < \ln(x+1) < x$ for $x > -1$, $x \neq 0$.

21.5. Since the graph is concave down, $y'(x)$ is strictly decreasing so that the slopes of the tangents and the secant satisfy $\frac{1}{a} > \frac{\ln b - \ln a}{b-a} > \frac{1}{b}$ (note that the slope of the secant is $y'(c)$ for some c in (a,b)).

22.1. $\ln(\sqrt[n]{n!}/n) = \ln(n!/n^n)^{1/n} = (1/n)\ln(\frac{1}{n} \cdot \frac{2}{n} \cdots \frac{n}{n}) = \sum_{k=1}^n \ln(k/n) \cdot (1/n)$.

22.2. The sum is a right endpoint Riemann sum with n subintervals on $[0,1]$ for $f(x) = \ln x$, so its limit as $n \to \infty$ is $\int_0^1 \ln x\,dx$.

22.3. $\int_0^1 \ln x\,dx = \lim_{t \to 0^+} \int_t^1 \ln x\,dx = \lim_{t \to 0^+}[x \ln x - x]_t^1$

$= \lim_{t \to 0^+}[-1 - t \ln t + t] = -1 - \lim_{t \to 0^+} \frac{\ln t}{1/t} = -1$.

22.4. Since $\lim_{n \to \infty} \ln(\sqrt[n]{n!}/n) = -1 = \ln(1/e)$, $\lim_{n \to \infty} \sqrt[n]{n!}/n = 1/e$.

23.1. $\int_a^b h(x)\,dx + ah(a) = bh(b) + \int_{h(b)}^{h(a)} h^{-1}(y)\,dy$, which is (23.1).

23.2. Let V denote the volume of the object described in the exercise. Then

$$V = \pi b[f(b)]^2 - \pi a[f(a)]^2 - 2\pi \int_{f(a)}^{f(b)} y f^{-1}(y)\,dy = V_{\text{shell}}.$$

24.1. Since $\pi[f(x) + tx]^2 \geq 0$ and $b - a > 0$, $\int_a^b \pi[f(x) + tx]^2\,dx \geq 0$.

24.2. $\int_a^b \pi[f(x)+tx]^2 dx = \int_a^b \pi[f(x)]^2 dx + t\int_a^b 2\pi x f(x) dx + t^2 \int_a^b x^2 dx$
$= V_{x\text{-axis}} + tV_{y\text{-axis}} + t^2(b^3-a^3)(b^3-a^3)3/3.$

24.3. If $At^2 + Bt + C \geq 0$, the quadratic $y = At^2 + Bt + C$ with $A > 0$ has at most one real root, hence $B^2 - 4AC \leq 0$.

24.4. $B^2 - 4AC \leq 0$ is equivalent to $V_{y\text{-axis}}^2 - 4V_{x\text{-axis}} \cdot (b^3-a^3)/3 \leq 0$, which is (24.2).

24.5. With $f(x) = mx$, $V_{x\text{-axis}} = \pi\int_a^b (mx)^2 dx = \pi m^2(b^3-a^3)/3$ and $V_{y\text{-axis}} = 2\pi\int_a^b mx^2 dx = 2\pi m(b^3-a^3)/3$, and so

$$V_{y\text{-axis}}^2 = \frac{4\pi^2 m^2}{9}(b^3-a^3)^2 = \frac{4\pi}{3}(b^3-a^3) \cdot \frac{\pi m^2}{3}(b^3-a^3)$$
$$= \frac{4\pi}{3}(b^3-a^3) \cdot V_{x\text{-axis}}.$$

24.6. When $t = -1$ (24.2) becomes $A-B+C \geq 0$ or $B-C \leq A$, which is $V_{y\text{-axis}} - V_{x\text{-axis}} \leq \pi(b^3-a^3)/3$. When $f(x) = x$, we have

$$\frac{2\pi}{3}(b^3-a^3) - \frac{\pi}{3}(b^3-a^3) = \frac{\pi}{3}(b^3-a^3),$$

so the inequality is best possible.

24.7. $0 \leq \int_a^b [f(x)+tg(x)]^2 dx = At^2 + Bt + C$ where $A = \int_a^b [g(x)]^2 dx$, $B = 2\int_a^b [f(x)g(x)]dx$, and $C = \int_a^b [f(x)]^2 dx$. As in Exercise 24.3, $B^2 \leq 4AC$, which is (24.3).

24.8. With $g(x) = 1$ (24.3) yields $A_R^2 \leq (V_{x\text{-axis}}/\pi)(b-a)$, so $V_{x\text{-axis}} \geq \pi A_R f_{\text{ave}}$. When $f(x) = k > 0$, $V_{x\text{-axis}} = \pi(b-a)k^2 = \pi \cdot (b-a)k \cdot k = \pi A_R f_{\text{ave}}$.

24.9. When $V_{x\text{-axis}} = 2\pi \bar{y} A_R$, the inequality in Exercise 24.8 becomes $2\pi \bar{y} A_R \geq \pi A_R f_{\text{ave}}$, or $\bar{y} \geq f_{\text{ave}}/2$.

25.1. Turn Figure 25.1 upside down.

26.1. When f is concave up, the error is minimized when the total area below the tangent line is maximized. This area is the base $b - a$ times the height h of the tangent line at the midpoint $x = (a+b)/2$, so it suffices to maximize h. That occurs when the point of tangency is at $x = (a+b)/2$.

26.2. See Figure S26.3, where we have rotated the horizontal line through the point P to a tangent line. The two shaded triangles have the same area.

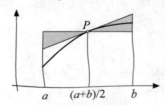

Figure S26.3.

Solutions to the Exercises

28.1. $\int_0^1 \frac{6dx}{\sqrt{4-x^2}} = 6\arcsin(x/2)|_0^1 = \pi$ and $\int_0^1 \frac{4dx}{1+x^2} = 4\arctan x|_0^1 = \pi$.

n	S_n for $\int_0^1 6dx/\sqrt{4-x^2}$	S_n for $\int_0^1 4dx/(1+x^2)$
10	3.1415955916	3.1415926139
20	3.1415928398	3.1415926529
40	3.1415926652	3.1415926535
100	3.1415926538	3.1415926535

28.2. Evaluating the integral suffices since the integrand is positive on $(0, 1)$, and thus so is the integral. Hence

$$0 < \int_0^1 \frac{x^4(1-x)^4}{1+x^2}dx = \int_0^1 \left(x^6 - 4x^5 + 5x^4 - 4x^2 + 4 - \frac{4}{1+x^2}\right)dx$$

$$= \frac{x^7}{7} - \frac{2x^6}{3} + x^5 - \frac{4x^3}{3} + 4x - 4\arctan x \Big|_0^1$$

$$= \frac{1}{7} - \frac{2}{3} + 1 - \frac{4}{3} + 4 - \pi = \frac{22}{7} - \pi.$$

29.1. Since f is concave up, (29.1) yields $\frac{2}{1+x} \leq \frac{\ln x}{x-1} \leq \frac{x+1}{2x}$. This is an improvement over the bounds from Exercises 21.1, as can be seen in Figure S29.1 (the solid black curve is the graph of f, the dashed curves the Hermite-Hadamard bounds, and the gray curves the bounds from Exercise 21.1).

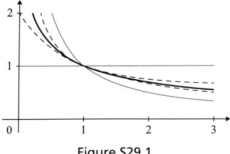

Figure S29.1

29.2. Using $a = n$ and $b = n+1$ in (29.3) yields $\sqrt{n(n+1)} \leq 1/\ln((n+1)) \leq (2n+1)/2$, and taking reciprocals yields (29.4).

30.1. The area $af(a)/2$ of the right triangle with base $[0, a]$ plus the area $\int_a^b f(x)dx$ under the curve minus the area $bf(b)/2$ of the right triangle with base $[0,b]$ equals A_{cart}.

30.2. $A_{\text{cart}} = \int_a^b f(x)dx + \frac{1}{2} xf(x)|_b^a = \frac{1}{2}\left[xf(x)|_b^a - 2\int_b^a f(x)dx\right]$.

30.3. $A_{\text{polar}} = \frac{1}{2}\int_\beta^\alpha [r(\theta)]^2 d\theta = \frac{1}{2}\int_\beta^\alpha [r(\theta)\cos\theta]^2 \sec^2\theta d\theta$

$$= \frac{1}{2}\int_\beta^\alpha [r(\theta)\cos\theta]^2 d(\tan\theta)$$

30.4. In the suggested change of variables, $d(f(x)/x) = d(\tan\theta)$ and $A_{\text{polar}} = \frac{1}{2}\int_b^a x^2 d(f(x)/x)$.

30.5. Using integration by parts with $u = x^2$ and $v = f(x)/x$ yields $A_{\text{polar}} = \frac{1}{2}[x^2 \frac{f(x)}{x}|_b^a - \int_b^a \frac{f(x)}{x} \cdot 2x dx] = \frac{1}{2}[xf(x)|_b^a - 2\int_b^a f(x)dx] = A_{\text{cart}}$.

30.6. Analogous to Exercise 30.1 we have $A_{\text{cart}} = \int_B^A g(y)dy + \frac{1}{2}bf(b) - \frac{1}{2}af(a)$. Adding this to the result in Exercise 30.1 and dividing by 2 yields $A_{\text{cart}} = \frac{1}{2}(\int_a^b f(x)dx + \int_B^A g(y)dy)$. By Exercise 30.5, this is also A_{polar}.

30.7. When $af(a) = bf(b)$ for all a and b (in the domain of f), then $xf(x)$ is a positive constant c, i.e., $f(x) = c/x$.

31.1. The graph of $r = \sec\theta$ is the line $x = 1$ in Cartesian coordinates. So for α in $[0, \pi/2)$, $\int_0^\alpha \sec^2\theta d\theta$ is twice the area of a right triangle with legs 1 and $\tan\alpha$, So $\int_0^\alpha \sec^2\theta d\theta = \tan\alpha$ and $\int \sec^2\theta d\theta = \tan\theta + C$.

31.2. The graph of $r = 1/(a\cos\theta + b\sin\theta)$ is the line $ax + by = 1$ in Cartesian coordinates. For α in $[0, \pi/2)$, the ray $\theta = \alpha$ intersects the line at the point with Cartesian coordinates $(1/(a + b\tan\alpha), 1/(a\cot\alpha + b))$. The integral equals twice the area of a triangle with base $1/a$ and altitude $1/(a\cot\alpha + b)$, so $\int_0^\alpha \frac{d\theta}{(a\cos\theta + b\sin\theta)^2} = \frac{1}{a(a\cot\alpha + b)}$, which establishes the result.

31.3. In Cartesian coordinates, $r = 1/(1 + \cos\theta)$ is $y^2 = 1 - 2x$, a parabola opening to the left with vertex $(1/2, 0)$ and y-intercepts $(0, \pm 1)$ as shown in Figure S31.3. The ray $\theta = \alpha$ intersects the parabola at the point P with Cartesian coordinates $(\cos\alpha/(1 + \cos\alpha), \sin\alpha/(1 + \cos\alpha))$.

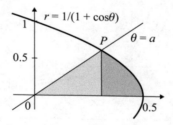

Figure S31.3

The area of the light gray triangle is $\sin\alpha\cos\alpha/[2(1 + \cos\alpha)^2]$, and the area of the dark gray parabolic region is $[\sin\alpha(1 - \cos\alpha)]/[3(1 + \cos\alpha)^2]$. Hence

$$\int_0^\alpha \frac{d\theta}{(1 + \cos\theta)^2} = \frac{\sin\alpha\cos\alpha}{(1 + \cos\alpha)^2} + \frac{2\sin\alpha(1 - \cos\alpha)}{3(1 + \cos\alpha)^2} = \frac{\sin\alpha(2 + \cos\alpha)}{3(1 + \cos\alpha)^2},$$

which establishes the result.

32.1. For a cylinder with base radius r and height h, $A_0 = A_m = A_1 = \pi r^2$ and (32.1) yields $V = (h/6)(6\pi r^2) = \pi r^2$, the correct volume. For a cone with base radius r and height h, $A_0 = \pi r^2$, $A_m = \pi r^2/4$, $A_1 = 0$, and (32.1) yields $V = (h/6)(2\pi r^2) = \pi r^2/3$, the correct volume.

Solutions to the Exercises

32.2. For the frustum in Figure 32.3b, $A_0 = b^2$, $A_m = [(a+b)/2]^2$, $A_1 = a^2$, and (32.1) yields $V = (h/6)(a^2 + (a+b)^2 + b^2) = (h/3)(a^2 + ab + b^2)$. So yes, the ancient formula is a special case of the prismoidal formula, and yes, it is exact since the frustum is a prismatoid.

Part III   Infinite Series

33.1. (a) $\frac{1}{4} + \frac{1}{16} + \frac{1}{64} + \cdots + \frac{1}{4^n} + \cdots = \frac{1}{3}$

(b) $\frac{2}{9} + \frac{2}{9}(\frac{1}{3}) + \frac{2}{9}(\frac{1}{3})^2 + \cdots + \frac{2}{9}(\frac{1}{3})^n + \cdots = \frac{1}{3}$.

33.2. Since $\triangle PQR$ and $\triangle PST$ are similar, $ST/PS = PQ/QR$ so that
$$\frac{1 + r + r^2 + r^3 + \cdots}{1} = \frac{1}{1-r}.$$

34.1. $1 + 2(\frac{1}{2}) + 3(\frac{1}{4}) + 4(\frac{1}{8}) + 5(\frac{1}{16}) + \cdots = 4$. It represents the derivative of $1 + x + x^2 + x^3 + \cdots = 1/(1-x)$ at $x = 1/2$.

35.1. Since $1 + 2 + 3 + \cdots + n = n(n+1)/2$, the nth term of the series is $2/n(n+1)$, and the nth partial sum is $2S_n$, where S_n is the nth partial sum in Example 35. Hence the sum of this series is $\lim_{n\to\infty} 2S_n = 2$.

36.1. Figure 36.2 illustrates how cubes with volumes $1, 1/8, \ldots, 1/n^3$ fit inside a box with volume $1 \cdot 1 \cdot 3/2 = 3/2$. Hence the sequence of partial sums of $\sum_{k=1}^{\infty} 1/k^3$ is increasing and bounded above, so the series converges.

36.2. Using (36.1) with $a = 1$, $b = n/(n+1)$, and $r = 1/(n+2)$ yields $[n/(n+1)]^{(n+1)/(n+2)} < (n+1)/(n+2)$ so that $[1 + (1/n)]^{(n+1)/(n+2)} > 1 + [1/(n+1)]$, hence $[1 + (1/n)]^{(n+1)} > [1 + 1/(n+1)]^{(n+2)}$ so that the sequence is decreasing. The terms of the sequence are positive, hence bounded below.

40.1. By counting squares in the grid, the tangent of the small acute angle in the dark gray triangle is $1/7$ while the tangent of the small acute angle in each of the light gray triangles $2/6 = 1/3$, The tangent of the union of those three angles is $5/5 = 1$, which proves (40.1).

40.2. The tangents of the three acute angles in the lower left corner of Figure 40.2 are $2/10 = 1/5$, $5/10 = 1/2$, and $1/8$. The tangent of their union is $8/8 = 1$, which proves (40.2). The successive partial sums π_n of the series
$$\pi = 4 \sum_{n=0}^{\infty} \frac{(-1)^n}{2n+1}[(1/2)^{2n+1} + (1/5)^{2n+1} + (1/8)^{2n+1}]$$
are

$$\pi_0 = 3.30$$
$$\pi_1 = 3.1200625$$
$$\pi_2 = 3.145342914\cdots$$
$$\vdots$$
$$\pi_{12} = 3.141592654\cdots$$
$$\pi_{13} = 3.141592653\cdots$$

so to eight decimal places, $\pi \approx 3.14159265$. While the convergence to π with Strassnitzky's formula is slower, it may have been more suited to hand calculation than Hutton's since the powers of $1/2$, $1/5$, and $1/8$ have terminating decimals while those of $1/3$ and $1/7$ do not.

40.3. Combining the hint with the area $\int_0^{1/4} \sqrt{x - x^2}\, dx$ under the semicircle over the interval $[0, 1/4]$ yields $\frac{\pi}{24} = \frac{\sqrt{3}}{32} + \int_0^{1/4} \sqrt{x - x^2}\, dx$.

40.4. $\sqrt{1 - x} = 1 - \frac{1}{2}x - \frac{1}{2^2}\frac{x^2}{2!} - \frac{1 \cdot 3}{2^3}\frac{x^3}{3!} - \frac{1 \cdot 3 \cdot 5}{2^4}\frac{x^4}{4!} - \cdots,$

$\sqrt{x}\sqrt{1 - x} = x^{1/2} - \frac{1}{2}x^{3/2} - \frac{1}{2^2}\frac{x^{5/2}}{2!} - \frac{1 \cdot 3}{2^3}\frac{x^{7/2}}{3!} - \frac{1 \cdot 3 \cdot 5}{2^4}\frac{x^{9/2}}{4!} - \cdots,$

$\int_0^{1/4} \sqrt{x}\sqrt{1 - x}\, dx = \frac{2}{3}\left(\frac{1}{4}\right)^{3/2} - \frac{1}{2} \cdot \frac{2}{5}\left(\frac{1}{4}\right)^{5/2} - \frac{1}{2^2} \cdot \frac{2}{7}\left(\frac{1}{4}\right)^{7/2}$

$\qquad - \frac{1 \cdot 3}{2^3} \cdot \frac{2}{9}\left(\frac{1}{4}\right)^{9/2} - \frac{1 \cdot 3 \cdot 5}{2^4} \cdot \frac{2}{11}\left(\frac{1}{4}\right)^{11/2} - \cdots$

$\qquad = \frac{1}{12} - \frac{1}{5 \cdot 2^5} - \frac{1}{4 \cdot 7 \cdot 2^7} - \frac{1 \cdot 3}{4 \cdot 6 \cdot 9 \cdot 2^9} - \frac{1 \cdot 3 \cdot 5}{4 \cdot 6 \cdot 8 \cdot 11 \cdot 2^{11}} - \cdots.$

Part IV Additional Topics

41.1. $u/2 = (1/2) \int_0^{\tan^{-1}(s/c)} \sec 2\theta\, d\theta = (1/4) \ln |\sec 2\theta + \tan 2\theta|_0^{\tan^{-1}(s/c)}$, but since $c^2 - s^2 = 1$, $\sec(2\tan^{-1}(s/c)) = c^2 + s^2$ and $\tan(2\tan^{-1}(s/c)) = 2cs$. Hence $u/2 = (u/4) \ln(c + s)^2 = (u/2) \ln(c + s)$, and so $u = \ln(c + s)$.

41.2. (a) $f(x) + g(x) = \frac{x^2 + 2x + 2}{x^4 + 1} = \frac{x^2 + 2x + 2}{(x^2 + 2x + 2)(x^2 - 2x + 2)} = \frac{1}{x^2 - 2x + 2}.$
(b) $h(-x) = f(-x) + g(-x) = f(x) - g(x)$, then solve this with $h(x) = f(x) + g(x)$ to yield $f(x) = [h(x) + h(-x)]/2$, $g(x) = [h(x) - h(-x)]/2$.
(d) when $h(x) = e^x$, $f(x) = \cosh x$ and $g(x) = \sinh x$.

42.1. $\frac{d}{du} \cosh u = \frac{d}{du} \sec(gd\, u) = \sec(gd\, u) \tan(gd\, u) \operatorname{sech} u = \sinh u$

$\frac{d}{du} \tanh u = \frac{d}{du} \sin(gd\, u) = \cos(gd\, u) \operatorname{sech} u = \operatorname{sech}^2 u$

$\frac{d}{du} \operatorname{sech} u = \frac{d}{du} \cos(gd\, u) = -\sin(gd\, u) \operatorname{sech} u = -\operatorname{sech} u \tanh u$

$\frac{d}{du} \operatorname{csch} u = \frac{d}{du} \cot(gd\, u) = -\csc^2(gd\, u) \operatorname{sech} u = -\operatorname{csch} u \coth u$

$\frac{d}{du} \coth u = \frac{d}{du} \csc(gd\, u) = -\csc(gd\, u) \cot(gd\, u) \operatorname{sech} u = -\operatorname{csch}^2 u.$

42.2. (a) $gd(-u) = \arctan(\sinh(-u)) = \arctan(-\sinh(u))$
$\qquad = -\arctan(\sinh(u)) = -gd(u).$
(b) $\lim_{u \to \infty} gd\, u = \lim_{u \to \infty} \arctan(\sinh u) = \lim_{x \to \infty} \arctan(x) = \pi/2.$
(c) $\lim_{u \to -\infty} gd\, u = \lim_{u \to -\infty} \arctan(\sinh u) = \lim_{x \to -\infty} \arctan(x) = -\pi/2.$

(d) $\sec(\text{gd } x) + \tan(\text{gd } x) = e^{\text{gd}^{-1}(\text{gd } x)} = e^x$.

(e) $\tan(\frac{1}{2}\text{gd}x) = \frac{1-\cos(\text{gd}x)}{\sin(\text{gd}x)} = \frac{1-\text{sech } x}{\tanh x} = \frac{\cosh x - 1}{\sinh x}$

$= \frac{2\sinh^2(x/2)}{2\sinh(x/2)\cosh(x/2)} = \tanh(x/2)$.

(f) $\tan(\frac{1}{2}\text{gd}x) = \tanh(x/2) = \frac{e^{x/2} - e^{-x/2}}{e^{x/2} + e^{-x/2}} = \frac{e^x - 1}{e^x + 1} = \tan(\arctan e^x - \frac{\pi}{4})$, and hence $\text{gd } x = 2\arctan e^x - (\pi/2)$.

44.1. Rotating the y- and z-axes through an angle θ ($0 \leq \theta \leq \pi/2$) as in this Cameo yields $x^2 + (\bar{y}\cos\theta - \bar{z}\sin\theta)^2 = r^2$ as the equation of the cylinder. Setting $\bar{z} = 0$ and replacing $\bar{y}$ by y yields $x^2 + (\cos^2\theta)y^2 = r^2$ as the equation of the intersection. This is a circle, an ellipse, or two parallel lines when $\theta = 0$, $0 < \theta < \pi/2$, or $\theta = \pi/2$, respectively.

45.1. From Example 45.1 we have $\sqrt[n]{n!} < (n+1)/2$, which is equivalent to the desired inequality.

Part V Appendix: Some Precalculus Topics

46.1. Let the parabola be given by $y = (1/4)x^2$. Then the points are $F(0, 1)$, $P(a, b)$ with $b = a^2/4$, and $Q(a, -1)$. $\triangle FPQ$ is isosceles since $|FP| = |PQ|$. The tangent line at P is $y = (a/2)[x - (a/2)]$, which intersects the x-axis (the tangent line at the vertex) at $R(a/2, 0)$. PQ is the line $y = 1 - (2/a)x$, which also passes through R. Since the slopes of PR and FQ are negative reciprocals, PR and FQ are perpendicular. Thus $\triangle FPR$ and $\triangle QPR$ are congruent right triangles, so PR bisects $\angle FPQ$.

47.1. (a) Use the Pythagorean theorem on the largest right triangle.
(b) The triangles with hypotenuses $\tan\theta$ and $\cot\theta$ are similar.

48.1. Evaluating the lengths of the legs of the gray right triangle in Figure 48.2 (note that the smaller acute angle has measure $\alpha - \beta$) yields

$$\sin(\alpha - \beta) = \sin\alpha\cos\beta - \cos\alpha\sin\beta,$$
$$\cos(\alpha - \beta) = \cos\alpha\cos\beta + \sin\alpha\sin\beta.$$

48.1. Divide all lengths in Figures 48.1b and 48.2 by $\cos\alpha\cos\beta$.

48.2. Replacing x by $x/2$ in two of the cosine formulas yields $\cos x = 2\cos^2(x/2) - 1$ and $\cos x = 1 - 2\sin^2(x/2)$. Solving for $\cos(x/2)$ and $\sin(x/2)$ yields the half angle formulas. The sign is positive in the sine formula if $x/2$ is in quadrant I or II, negative if $x/2$ is in quadrant III or IV. The sign is positive in the cosine formula if $x/2$ is in quadrant I or IV, negative if $x/2$ is in quadrant II or III.

49.2. Evaluating the lengths of the legs of the right triangle with an acute angle measuring $3x$ and employing the double angle cosine formulas yields

$$\sin 3x = 2\sin x \cos 2x + \sin x = 2\sin x(1 - 2\sin^2 x) + \sin x = 3\sin x - 4\sin^3 x,$$
$$\cos 3x = 2\cos x \cos 2x - \cos x = 2\cos x(2\cos^2 x - 1) - \cos x = 4\cos^3 x - 3\cos x.$$

50.1. (a) $x^6 + 1 = (x^2 + 1)(x^4 - x^2 + 1)$
$= (x^2 + 1)[(x^4 + 2x^2 + 1) - 3x^2]$
$= (x^2 + 1)[(x^2 + 1)^2 - (\sqrt{3}x)^2]$
$= (x^2 + 1)(x^2 - \sqrt{3}x + 1)(x^2 + \sqrt{3}x + 1).$

(b) $x^8 + x^4 + 1 = (x^8 + 2x^4 + 1) - x^4$
$= (x^4 + 1)^2 - (x^2)^2$
$= (x^4 + x^2 + 1)(x^4 - x^2 + 1)$
$= [(x^4 + 2x^2 + 1) - x^2][(x^4 + 2x^2 + 1) - 3x^2]$
$= [(x^2 + 1)^2 - x^2][(x^2 + 1)^2 - (\sqrt{3}x)^2]$
$= (x^2 - x + 1)(x^2 + x + 1)(x^2 - \sqrt{3}x + 1)(x^2 + \sqrt{3}x + 1).$

References

C. Alsina and R. B. Nelsen, "Teaching tip: The limit of (sin$t$)/t," *College Mathematics Journal*, **41** (2010), p. 192.

———— "Proof without words: The triple angle sine and cosine formulas," *Mathematics Magazine*, **85** (2012), p. 43.

M. K. Brozinsky, "Proof without words," *College Mathematics Journal*, **25** (1994), p. 98.

F. Burk, "Behold! The midpoint rule is better than the trapezoidal rule for concave functions," *College Mathematics Journal*, **16** (1985), p. 56.

R. D. Carmichael, "On the representation of the trigonometric functions by lines," *American Mathematical Monthly*, **15** (1908), pp. 199–200.

R. Courant, *Differential and Integral Calculus*, Vol. 1, Interscience, New York, 1937.

D. Cruz-Uribe, "The relation between the root and ratio tests," *Mathematics Magazine*, **70** (1997), pp. 214–215.

M. R. Cullen, "Cylinder and cone cutting," *College Mathematics Journal*, **28** (1997), pp. 122–123.

A. Cupillari, "Proof without words: $1^3 + 2^3 + \cdots + n^3 = (n(n+1))^2/4$" *Mathematics Magazine*, **62** (1989), p. 259.

P. Deiermann, "The method of last resort (Weierstrass substitution)", *College Mathematics Journal*, **29** (1998), p. 17.

H. Dörrie, *100 Great Problems of Elementary Mathematics*, Dover Publications, New York, 1965.

W. Dunham, *Journey Through Genius: The Great Theorems of Mathematics*, John Wiley & Sons, Inc., New York, 1990.

R. Euler, "A note on differentiation," *College Mathematics Journal*, **17** (1986), pp. 166–167.

H. Eves, *An Introduction to the History of Mathematics, Fifth Edition*, Saunders College Publishing Company, Philadelphia, 1983.

G. Fredricks and R. B. Nelsen, "Summation by parts," *College Mathematics Journal*, **23** (1992), pp. 39–42.

N. A. Friedman, "A picture for the derivative," *American Mathematical Monthly*, **84** (1977), pp. 470–471.

M. Gardner, "Mathematical Games," *Scientific American*, **229** (1973), p. 115.

S. W. Golomb, "A geometric proof of a famous identity," *Mathematical Gazette*, **49** (1965), pp. 198–200.

R. N. Greenwell, "Why Simpson's rule gives exact answers for cubics," *Mathematical Gazette*, **83** (1999), p. 508.

R. H. Hammack and D. W. Lyons, "Proof without words," *College Mathematics Journal*, **36** (2005), p. 72.

——— "Alternating series convergence: a visual proof," *Teaching Mathematics and Its Applications*, **25** (2006), pp. 58–60.

D. Hartig, "On the differentiation formula for $\sin\theta$," *American Mathematical Monthly*, **96** (1989), p. 252.

M. Hudelson, "Proof without words: The alternating harmonic series sums to $\ln 2$," *Mathematics Magazine*, **83** (2010), p. 294.

D. Kalman, "$(1 + 2 + \cdots + n)(2n + 1) = 3(1^2 + 2^2 + \cdots + n^2)$", *College Mathematics Journal*, **22** (1991), p. 124.

E. Key, "Disks, shells, and integrals of inverse functions," *College Mathematics Journal*, **25** (1994), pp. 136–138.

G. Kimble, "Euler's other proof," *Mathematics Magazine*, **60** (1987), p. 282.

M. K. Kinyon, "Another look at some p-series," *College Mathematics Journal*, **37** (2006), pp. 385–386.

S. H. Kung, "Proof without words: The Weierstrass substitution," *Mathematics Magazine*, **74** (2001), p. 393.

W. Lushbaugh, *Mathematical Gazette*, **49** (1965), p. 200.

J. H. Mathews, "The sum is one," *College Mathematics Journal*, **22** (1991), p. 322.

N. S. Mendelsohn, "An application of a famous inequality," *American Mathematical Monthly*, **58** (1951), p. 563.

K. Menger, *Calculus: A Modern Approach*, Ginn, Boston, 1955 (reprinted by Dover Publications, Inc, Mineola, NY, 2007).

C. C. Mumma II, "$N!$ and the root test," *American Mathematical Monthly*, **93** (1986), p. 561.

R. B. Nelsen, "Proof without words: The substitution to make a rational function of the sine and cosine," *Mathematics Magazine*, **62** (1989), p. 267.

———, "Napier's inequality (two proofs)," *College Mathematics Journal*, **24** (1993), p. 165.

———, "Symmetry and integration," *College Mathematics Journal*, **26** (1995), pp. 39–41.

———, "One figure, six identities," *College Mathematics Journal*, **31** (2000), pp. 145–146.

———, "Proof without words: Steiner's problem on the number e," *Mathematics Magazine*, **82** (2009), p. 102.

I. Niven, "Which is larger, e^π or π^e?" *Two-Year College Mathematics Journal*, **3** (1972), pp. 13–15.

R. Paré, "A visual proof of Eddy and Fritsch's minimal area property," *College Mathematics Journal*, **26** (1995), pp. 43–44.

J. M. H. Peters, "The Gudermannian," *Mathematical Gazette*, **68** (1984), pp. 192–196.

I. Richards, "Proof without words: Sum of integers'" *Mathematics Magazine*, **57** (1984), p. 104.

N. Schaumberger, "An alternate classroom proof of the familiar limit for e, *Two-Year College Mathematics Journal*, **3** (1972), pp. 72–73.

———, "The derivatives of $\text{arcsec} x$, $\text{arctan} x$, and $\tan x$," *College Mathematics Journal*, **17** (1986), pp. 244–246.

———, "A coordinate approach to the AM-GM inequality," *Mathematics Magazine*, **64** (1991), p. 273

M.-K. Siu, "Proof without words: Sum of squares," *Mathematics Magazine*, **57** (1984), p. 92.

C. G. Spaht and C. M. Johnson, "Mathematics without words," *College Mathematics Journal*, **32** (2001), p. 109.

M. R. Spiegel, "On the derivatives of trigonometric functions," *American Mathematical Monthly*, **63** (1956), pp. 118–120.

M. Spivak, *Calculus*, Cambridge University Press, Cambridge, 2006.

S. Sridharma, "The derivative of $\sin\theta$," *College Mathematics Journal*, **30** (1999), pp. 314–315.

G. Strang, "Polar area is the average of strip areas," *American Mathematical Monthly*, **100** (1993), pp. 250–254.

J. Tong, "Area and perimeter, volume and surface area," *College Mathematics Journal*, **28** (1997), p. 57.

H. Unal, "Proof without words: Sum of an infinite series," *College Mathematics Journal*, **40** (2009), p. 39.

The Viewpoints 2000 Group, "Proof without words: Geometric series," *Mathematics Magazine*, **74** (2001), p. 320.

R. Woods, "The trigonometric functions of half or double an angle," *American Mathematical Monthly*, **43** (1936), pp. 174–175.

W. Zimmerman and S. Cunningham (editors), *Visualization in Teaching and Learning Mathematics*, Mathematical Association of America, Washington, 1991.

Index

NOTE: Numbers refer to Cameos rather than pages.

alternating harmonic series 38
alternating series test 39
annulus 14
approximating π 2, 28, 40
area, derivatives of 14
   integral 9
   of a circular sector 1
   of a regular polygon 2
   polar 30, 31
Aristarchus's inequalities 11
arithmetic mean 10–12, 15, 21, 29, 45
arithmetic mean-geometric mean inequality 11, 15, 29, 45
average value of a function 24, 29

Bernoulli's inequality 11

Cauchy-Schwarz inequality 24
centroid 24
chain rule 6
circle 1, 2, 14, 43, 44
circular sector 1
combinatorial identities 16
completing the square 44, 50
composite function 6
cone 15, 32, 43, 44
conic sections 43, 44
cubes, sums of 16, 17
cylinder 14, 32, 43, 44

derivatives 3
   and polar area 31
   of a composite function 6
   of a product 4
   of a quotient 5
   of a square 4
   of area 14

   of hyperbolic functions 42
   of the arcsine 9
   of the arctangent 8
   of the cosine 7
   of the sine 7
   of the tangent 8
   of volume 14
directrix 43, 46
disk method 23, 32
double angle formulas 31, 49

e as a limit 12, 21, 22, 36, 45
   to the π power 13
eccentricity 43
ellipse 43, 44
equilateral triangle 14
Euler-Mascheroni constant 37
even part of a function 41
exponential function 12, 13

focus 43, 46
frustum of a pyramid 32
fundamental theorem 9, 31

geometric mean 10–12, 15, 21, 29, 45
geometric series 33
   differentiation of 34
   partial sums of 17, 33
Gudermannian 42

half angle formulas 2, 49
harmonic series 37
   alternating 38
Hermite-Hadamard inequality 29
Hutton's formula 40
hyperbola 43, 44
hyperbolic functions 41, 42

identric mean 10
inequalities, Aristarchus's 11
　　for a rectangle 15
　　for a rectangular box 15
　　for e 12, 21, 29, 36
　　tangent line 11
inequality, arithmetic mean-geometric mean
　　　(AM-GM) 11, 15, 45
　　Bernoulli's 11
　　between a^b and b^a 13
　　between e^π and π^e 13
　　between 22/7 and π 28
　　Cauchy-Schwarz 24
　　Hermite-Hadamard 29
　　Mengoli's 37
　　Napier's 21
　　Weighted AM-GM 11, 36
integral test 36, 37
integration and symmetry 20
　　by parts 18, 23, 31
　　numeric 25–28

limit for e 12, 21, 22, 36, 45
　　for the natural logarithm 11
　　of $(\sin t)/t$ 1
　　of the nth root of $n!$ 22, 45
logarithmic mean 10, 21, 29
　　differentiation 13

Maclaurin series 40
mathematical induction 16
mean, arithmetic 10–12, 15, 21, 29, 45
　　geometric 10–12, 15, 21, 29, 45
　　identric 10
　　logarithmic 10, 21, 29
mean value theorem 10
means 10–12, 15, 21, 29, 45
Mengoli's inequality 37
midpoint rule 12, 20, 25, 26, 29
monotone sequence theorem 36, 37

n factorial 22, 45
Napier's inequality 21
natural logarithm as a limit 11
nth root of $n!$ 22, 45
numeric integration 25–28

odd numbers, sums of 17

odd part of a function 41
optimization 15

π, approximating 2, 28, 40
　　to the e power 13
parabola 43, 44, 46
partial fractions 19, 50
partial sums of alternating harmonic series
　　　38
　　of alternating series 39
　　of geometric series 17, 33
　　of harmonic series 37
　　of telescoping series 17, 35
perimeter 2, 14
polar area 30, 31, 41
positive integers, sums of 16, 17
prismatoid 32
prismoidal formula 32
product rule 4
Putnam Competition 11, 20, 28, 45
pyramid 15, 32
　　frustum of 32

quotient rule 5

ratio test 45
reciprocal rule 5
Regiomontanus's maximum problem 15
Riemann sums 16, 22, 23
root test 45
rule, chain 6
　　midpoint 12, 20, 25, 26, 29
　　product 4
　　reciprocal 5
　　quotient 5
　　Simpson's 27, 32
　　trapezoidal 12, 20, 25, 29

secant lines 3, 6, 10, 21
shell method 23
series, geometric 33
　　harmonic 36
　　Maclaurin 40
　　telescoping 35
Simpson's rule 27, 32
slope axis 3
solids of revolution 23, 24
sphere 14, 32, 43

Index

squares, sums of 16, 17
Strassnitzky's formula 40
substitution, Weierstrass 19
summation by parts 17
sums of cubes 16, 17
  of odd numbers 17
  of positive integers 16, 17
  of squares 16, 17
  of triangular numbers 17
  Riemann 16, 22, 23
surface area 14
symmetry and integration 20

tangent lines 3, 10–12, 46

telescoping series 35
  partial sums of 17, 35
torus 14
trapezoidal rule 12, 20, 25, 29
triangle, equilateral 14
triangular numbers 16, 17
triple angle formulas 49

volume by disks 23
  by shells 23
  derivatives of 14

Weierstrass substitution 19
weighted AM-GM inequality 11, 36

About the Author

Roger B. Nelsen was born in Chicago, Illinois. He received his B.A. in mathematics from DePauw University in 1964 and his Ph.D. in mathematics from Duke University in 1969. Roger was elected to Phi Beta Kappa and Sigma Xi, and taught mathematics and statistics at Lewis & Clark College for forty years before his retirement in 2009. His previous books include *Proofs Without Words*, MAA 1993; *An Introduction to Copulas*, Springer, 1999 (2nd ed. 2006); *Proofs Without Words II*, MAA, 2000; *Math Made Visual* (with Claudi Alsina), MAA, 2006; *When Less Is More* (with Claudi Alsina), MAA, 2009; *Charming Proofs* (with Claudi Alsina), MAA, 2010; *The Calculus Collection* (with Caren Diefenderfer), MAA, 2010; *Icons of Mathematics* (with Claudi Alsina), MAA, 2011, *College Calculus* (with Michael Boardman), MAA, 2015, and *A Mathematical Space Odyssey* (with Claudi Alsina), MAA, 2015.